Pro Power BI Theme Creation

JSON Stylesheets for Automated Dashboard Formatting

Second Edition

Adam Aspin

Apress®

Pro Power BI Theme Creation: JSON Stylesheets for Automated Dashboard Formatting

Adam Aspin
STAFFORD, UK

ISBN-13 (pbk): 978-1-4842-9632-5   ISBN-13 (electronic): 978-1-4842-9633-2
https://doi.org/10.1007/978-1-4842-9633-2

Managing Director, Apress Media LLC: Welmoed Spahr
Acquisitions Editor: Mark Powers
Development Editor: Laura Berendson

Cover designed by eStudioCalamar

Cover image by Piro on Pixabay (www.pixabay.com)

Distributed to the book trade worldwide by Apress Media, LLC, 1 New York Plaza, New York, NY 10004, U.S.A. Phone 1-800-SPRINGER, fax (201) 348-4505, e-mail orders-ny@springer-sbm.com, or visit www.springeronline.com. Apress Media, LLC is a California LLC and the sole member (owner) is Springer Science + Business Media Finance Inc (SSBM Finance Inc). SSBM Finance Inc is a **Delaware** corporation.

For information on translations, please e-mail booktranslations@springernature.com; for reprint, paperback, or audio rights, please e-mail bookpermissions@springernature.com.

Apress titles may be purchased in bulk for academic, corporate, or promotional use. eBook versions and licenses are also available for most titles. For more information, reference our Print and eBook Bulk Sales web page at http://www.apress.com/bulk-sales.

Any source code or other supplementary material referenced by the author in this book is available to readers on GitHub (https://github.com/Apress). For more detailed information, please visit http://www.apress.com/source-code.

Paper in this product is recyclable

Table of Contents

About the Author

Adam Aspin is an independent business intelligence consultant based in the United Kingdom. He has worked in business intelligence and analytics for over 25 years and now focuses on Power BI. During this time, he has developed several dozen BI and analytics systems based on the Microsoft data platform. Adam has been creating JSON theme files since the feature was first introduced in Power BI Desktop and has delivered corporate Power BI themes for dozens of clients across Europe.

Adam is a graduate of Oxford University. He has applied his skills for a range of clients in finance, banking, utilities, telecoms, construction, and retail. He is the author of a number of Apress books: *Pro Power BI Dashboard Creation*; *Pro DAX and Data Modeling in Power BI*; *Pro Data Mashup for Power BI*; *SQL Server 2012 Data Integration Recipes*; *Business Intelligence with SQL Server Reporting Services*; *High Impact Data Visualization in Excel with Power View, 3D Maps, Get & Transform and Power BI*; and *Data Mashup with Microsoft Excel Using Power Query and M*.

About the Technical Reviewer

Natalie Jayne Heger is a business intelligence developer/consultant based in the West Midlands of England.

She has worked in the IT industry for over 20 years. She started professional development using VB6 and Microsoft Access in the late 1990s until discovering SQL Server was a far more robust database solution and worked with DTS/SSIS, SSRS, and SSAS for many years for a broad range of companies.

She now works in the construction industry developing solutions with Power BI and enjoys thinking outside the box in producing quality reporting solutions, particularly in relation to D365 integration.

Acknowledgments

Writing a technical book can prove to be a daunting challenge. So I am all the more grateful for all the help and encouragement that I have received from so many friends and colleagues.

First, my heartfelt thanks go to Mark Powers, the commissioning editor of this book. Throughout the publication process, Mark has been an exemplary mentor. He has provided much valuable guidance during the preparation of this book.

My deepest gratitude goes yet again to the Apress production team for managing this, our sixth book together, through the rocks and shoals of the publication process.

When delving into the arcane depths of technical products, it is all too easy to lose sight of the main objectives of a book. Fortunately, my good friend and former colleague Natalie Heger, the technical reviewer, has worked unstintingly to help me retain focus on the objectives of this book. She has also shared her considerable experience of Power BI Desktop in the enterprise and has helped me immensely with her comments and suggestions.

Finally, my deepest gratitude has to be reserved for the two people who have given the most to this book. They are my wife and son, who have always encouraged me to persevere while providing all the support and encouragement that anyone could want. I am very lucky to have both of them.

Introduction

As a Power BI user, you are used to delivering eye-catching analytical dashboards. However, formatting and reformatting dozens or even hundreds of visuals across a set of dashboards can be a time-consuming and needlessly repetitive task. This effort can require heroic levels of concentration and patience when you are faced with the task of creating a unified look and feel for a suite of reports – or standardizing an enterprise-wide analytics suite in Power BI.

This book is about helping you to minimize the time and effort you need to spend on formatting and tweaking dashboard visuals. It shows you how a small investment in time up front can save you hundreds of hours later when it comes to delivering analytics. All this is done by defining a theme file that can be used to specify – once and for all – the presentation of each and every visual that is built into Power BI Desktop.

Applying theme files makes your dashboards adopt instantly the look and feel that you have designed. Not only that, but any changes that you make later to the theme can be reapplied in a couple of clicks to update the presentation to match the latest definition. This is not only time saved, it is also a liberation from the drudgery of manual formatting. This means that you are now free to concentrate on delivering powerful analytics and data journeys that entrance and motivate your users.

Winning back this wasted time requires only a small up-front investment. You have to learn how the format pane in Power BI Desktop maps to JSON elements in a theme file. Then you discover how the theme file is structured to define visual formatting. In case you are not a JSON expert, this book also introduces you to the core elements of JSON that you need to understand in order to create powerful theme files.

You can, if you wish, read this book from start to finish, as it is designed to be a progressive self-tutorial. However, as Power BI Desktop themes can be applied at several levels of complexity (requiring progressively more advanced skills), the book is broken down into several separate chapters that correspond to the various levels and methods of theme generation. They are as follows.

Chapter 1 provides a high-level overview of what a theme file can do to accelerate dashboard production.

Chapter 2 explains how to create theme files directly inside Power BI Desktop. This serves as a basic introduction to how themes work in practice and how you can develop usable JSON files without writing a line of JSON code.

Chapter 3 shows how to create high-level theme files directly in JSON using theme classes to simplify theme file creation. This can avoid theme developers having to go into all the intricate layers of detail needed to apply visual formatting while producing effective themes quickly and easily.

Chapter 4 describes the techniques required for generic theme definition. These are theme files that can set common attributes of multiple visuals. This approach can help theme creators produce simple yet powerful theme files in a relatively short time.

Chapters 5 through 12 are a complete reference that explains each and every attribute of all the visuals currently available in Power BI Desktop. These chapters explain each detail of how every standard visual can be formatted in a JSON file. Every option and selection is explained so that you can tweak dashboard presentation down to the finest and most intricate level of detail.

Chapter 13 extends your knowledge by introducing the concept of cascading style definitions. These allow you to specify formatting attributes that apply at selected levels of formatting. This approach can make the maintenance of theme files considerably easier.

This book comes with several dozen theme files to help you both learn about themes in Power BI and to help you create your own theme files. Some of these files cover a specific aspect of formatting a dashboard, some apply to a single visual so that you can delve into the intricacies of a specific object, and some are complete and detailed report theme files containing several thousand lines of JSON.

This book also comes with a small sample data set that you can use to create visuals. As the focus is on form and not analysis, I prefer to use an extremely simplistic data structure so that the reader is free to concentrate on formatting and not the data itself.

The details on how to access and download the sample files are given in Appendix A.

Inevitably, not every question about styling dashboards can be answered and not every issue can be resolved in one book. I truly hope that I have answered many of the essential Power BI stylesheet questions that you will face and have provided ways of solving most of the pre-formatting challenges that you may encounter. I wish you good luck in using Power BI Desktop when delivering your insights. And I sincerely hope that you have as much fun with themes as I had writing this book.

—Adam Aspin

CHAPTER 1

■ ■ ■

Introduction to Power BI Themes

Power BI has taken the world by storm when it comes to creating attention-grabbing dashboards that empower users. It has come to dominate the analytics arena with its ease of use, wide range of connectivity options, and the variety of available visuals.

However, formatting (and reformatting) dashboard visuals can prove time-consuming and repetitive – as can standardizing the presentation of multiple dashboards to create a unified look and feel for a suite of reports. Most users would rather spend their time analyzing and delivering meaningful insights as opposed to applying colors and font choices to charts and tables.

This is where the creation and application of Power BI *themes* comes in. A theme is a standardized definition of some – or all – of the formatting of a dashboard. This can range from defining a color palette and a selection of font choices to the detailed specification of each and every type of built-in visual. Applying a theme allows you to format virtually every visual in a dashboard *instantly*. What is more, any changes that you subsequently make to a theme can be reapplied in a few clicks to update your dashboard's presentation. Themes can be created once, then applied to dozens or even hundreds of Power BI dashboards to guarantee a coherent and rigorously standardized presentation style across a department or even an entire organization.

Themes can be as simple – or as detailed – as you require. They can specify standardized fonts, text colors, and font sizes or set the tiniest details of how each individual type of visual is formatted. Indeed, themes are designed to work at multiple levels – starting with general elements and then progressing deeper and deeper into tightly specified definitions of the presentation of each element in a dashboard. This gives you a range of options when it comes to defining how you standardize Power BI dashboards.

Theme Basics

Themes are, technically, JSON files. That is, they use the JSON (JavaScript Object Notation) format to store the definition of how dashboard objects are formatted. But don't worry if you have never used – or even heard of – JSON before now, as I will be explaining all you need to know about using it in theme files during the course of this book. The only thing to retain for the moment is that themes can be thought of as external stylesheets for Power BI (very like the stylesheets that are used by Microsoft Word, for instance). Yet another advantage of defining formats with external files is that you can simply send your colleagues themes as small text-based files that they can then apply in a few clicks to any of their own Power BI dashboards. Themes can also be integrated into Power BI templates to accelerate even further the dashboard creation process.

Themes are applied to Power BI Desktop files. Once loaded, they become an integral part of the Power BI Desktop .pbix file. The formatting is then, of course, carried over into the Power BI service when you publish the dashboard. Moreover, you can switch between different theme files in a few clicks to try out different formatting approaches or apply a completely different look and feel. Equally fortunately, there is

A. Aspin, *Pro Power BI Theme Creation*, https://doi.org/10.1007/978-1-4842-9633-2_1

nothing definitive about applying a theme file, as you can switch to another Power BI presentation style based on a different theme at any time. So you are free to experiment with different styles until you find the look and feel that suits the particular dashboard that you are creating.

A theme allows you to preset virtually anything that is defined in the Power BI format pane. Consequently, it can contain formatting definitions for

- The color palette

- The page

- Inserted objects (text boxes, shapes, buttons, and images)

- Built-in visuals (tables, charts, maps, and other visuals)

You can even apply a theme to the Power BI Desktop Filter pane. The principle is simple: if you can alter presentation in the format pane of Power BI Desktop, then you can define the format using a theme file.

The portable nature of dashboard themes means that there are now a wide range of different themes that are freely available to enhance the presentation of Power BI dashboards. You may even have experimented with some of the themes that are available as an integral part of Power BI Desktop or shared through the Web.

After reading this book and learning how these themes are created, you, too, will be able to develop themes that you can share – with friends, colleagues, or even the world.

Anatomy of a Theme

It is important to understand from the start that themes operate at two distinct levels. Firstly, themes can be defined to operate in a "generic" way. That is, you can define a set of high-level specifications to set core aspects of a dashboard and the visuals it contains without going into low levels of detail. This way, you can quickly and easily define

- Colors

- Standard font attributes

- Shared elements for all visuals

- Certain elements specific to certain types of visuals

You will discover how to create "generic" themes in Chapter 3.

Then, at a more detailed level (and once you have acquired a thorough understanding of how themes work), you can specify the *exact formatting for each and every element of each visual* to give you total control of the appearance of each element in a dashboard. An in-depth definition of the JSON that is required at the level of individual visuals is the subject of Chapters 5 through 12.

However, don't get the idea that you have to choose between two different approaches. The truth is that a JSON theme file can "mix and match" the two approaches so that you can both define a set of overall style principles *and* then add highly specific definitions of the way that you want certain visuals to be displayed in a dashboard. The principle will always be that a more generic level of formatting applies unless a more detailed definition is found in the theme file.

Themes and the Format Pane

A theme is nothing more than the definition of the formatting that you would otherwise apply manually using the Power BI format pane. It follows that you need to know what the format pane is and how it is used to format dashboard objects. I will not be explaining the intricacies of the Format Pane in this book, however, and refer you to my book *Pro Power BI Dashboard Creation* for full details on how to format dashboards manually.

Nonetheless, it is worth reminding yourself of the core elements in the Format Pane and how formatting works at a generic level.

Essentially, formatting is a hierarchy of three elements:

- *Visual* (a table or chart – although this could be the page as well).

- *Element*: This is the aspect of a visual that you want to format. This corresponds to the expandable sections and subsections of the format pane in Power BI Desktop (they are often also called tabs or cards). The X axis section for a chart is an example of this.

- *Property* (often referred to as an attribute): This is the specific setting you define. This could be the font color, for instance.

You can see this (for a stacked bar chart visual) in Figure 1-1, where the Data labels element is expanded to show all the available properties that you can set for data labels.

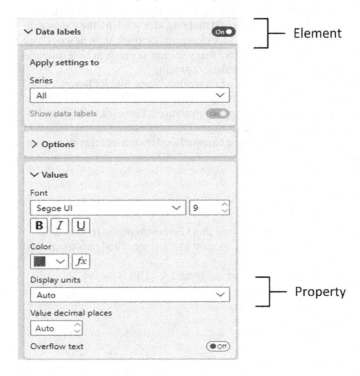

Figure 1-1. *The Format Pane*

It is important to bear this hierarchy in mind when creating theme files, as the entire JSON structure is predicated on this nested approach – as will be explained further in Chapter 3.

Writing JSON Files

As you will have to write or adapt actual JSON files for all but the simplest theme files, you will need to choose a tool to carry out your modifications. As JSON is a text-based format, there is a plethora of tools that are available to do this. It is entirely up to you to select a tool that you are happy with. A few that I have used are

- Notepad

- Notepad++

- Visual Studio

- Visual Studio Code

- An online JSON editor

There are literally dozens of tools available, and the choice of an application to use is entirely up to you. My only advice is to find something that you are comfortable with and that provides you with the balance of usability and power that you require. Above all, it must be able to save or export text files. As, in practice, you are likely to test and revise your choices over and over again, the tool that you choose needs to have multiple levels of undo and redo as well as having a powerful search and replace capability.

A tool that displays JSON in an easy-to-read format is a must when writing theme files for Power BI. This is because the JSON structure used by Power BI Desktop can get intricate, and consequently it helps considerably to be able to see the nested levels of text and how they are structured. However, I don't want to get too far ahead, yet, so will come back to this later in the book.

If you have not yet made up your mind, I suggest downloading Notepad++. This is a free text editor that can display JSON in a very readable format.

Applying Themes

Before looking at how you create your own themes, it is probably best that I demonstrate the far-reaching effects that themes can deliver. Hopefully, this will incentivize you to want to learn more and encourage you how to create your own theme files.

Take, for instance, a very simple dashboard like the one shown in Figure 1-2. (This is the file SampleDashboard.pbix from the sample files.)

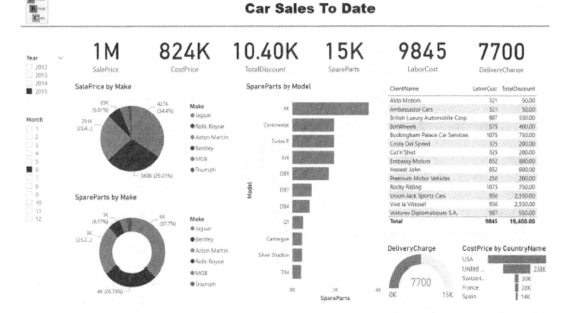

Figure 1-2. *An unformatted dashboard*

This dashboard is strictly "plain vanilla." That is, no formatting has been applied at all. Imagine that this dashboard contained 15 pages and that you have to make it not only presentable but also mapping to a corporate standard. This could take hours in a worst-case scenario.

However, you are fortunate enough to have a JSON theme file available that defines all the corporate standards. To apply it, all you have to do is

1. Click View to activate the View ribbon.

2. Click the Themes popup.

3. Click Browse for themes.

4. Navigate to the sample files and select the theme file (FullTheme.json in this example).

5. Click Open. After a few seconds, you will see the dialog in Figure 1-3.

Figure 1-3. *Successful theme file import*

6. Click Close.

That is it! Your dashboard (every object on every page) is now formatted, instantly. The result is shown in Figure 1-4.

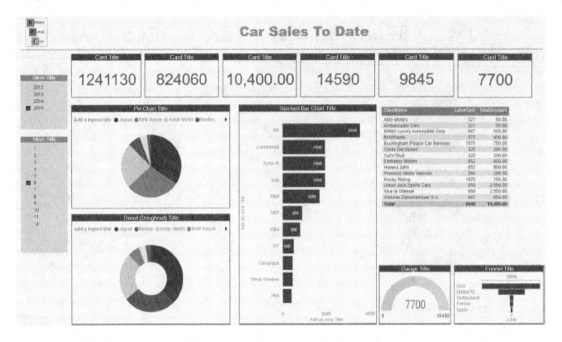

Figure 1-4. A dashboard after applying a theme file

You can now add any final formatting that you require to override the theme at the level of individual elements (a theme, after all, is a time-saver – not a straitjacket). The end result is almost certainly that you have saved a vast amount of time adding presentational touches and have ensured that your dashboard conforms to a standard look and feel.

▓ **Note**   The theme file that you applied in this example is not designed to show off subtleties of dashboard design – nor is it based on cutting-edge design principles. It is, however, designed to illustrate how quick and easy it is to reformat an entire dashboard using a theme file.

Conclusion

In this chapter, you have seen what themes are all about and exactly how powerful they can be. It is time, now, to move on to creating your own themes. Initially, we will begin by using Power BI Desktop to write the JSON for you – as you will discover in the next chapter.

CHAPTER 2

■ ■ ■

Create and Customize a Theme in Power BI Desktop

If you do not come from an IT background (or have never used JSON files before), and the thought of handcrafting complex text structures fills you with dread, then now is the time to relax and breathe deeply, because it is ridiculously easy to define a simple yet powerful Power BI theme file *without writing or modifying a single line of code*.

This is because Power BI Desktop allows you to specify (in a few clicks) a core subset of formatting attributes that will apply to your entire dashboard. Not only that, but once you have defined these attributes, Power BI Desktop will write the JSON theme file for you so that you can use it in other dashboards – or share with colleagues. All this is done using the *Customize current theme* option of the Themes button in the View menu.

While this approach does not let you craft every precise formatting attribute of each individual type of visual, it does allow you to create a standard presentation style in a few minutes at most – and you can subsequently take this theme as a basis for further development. What is more, the resulting theme file is an excellent introduction to how JSON themes function in Power BI.

Be aware, nonetheless, that creating themes using the Power BI Desktop interface does *not* allow you to attain the sophisticated levels of presentation that you can deliver using "handcrafted" JSON. So it is best to consider the techniques that you will discover in this chapter both as a practical introduction to the subject of themes and also as an extremely fast way of developing simple style principles that can be applied to entire dashboards.

To provide a little structure to the process, this "interactive theme generation" directly inside Power BI Desktop breaks down the theme elements that you can alter into the following five categories:

- *Name and colors*: Where you can give your theme a name and specify the color palette and key colors to use in a dashboard

- *Text*: Where you can define core text attributes such as font, size, and color for various elements in a dashboard

- *Visuals*: Where you can specify key common attributes of all visuals

- *Page*: Where you can set basic attributes of the dashboard pages

- *Filter pane*: That allows you to specify the way that the filter pane appears

You do not have to specify every option that Power BI Desktop makes available in the Customize current theme dialogs. You can change a couple of options – or reset everything that is available (and all steps in between).

© Adam Aspin 2023
A. Aspin, *Pro Power BI Theme Creation*, https://doi.org/10.1007/978-1-4842-9633-2_2

Creating a Simple Theme

To get you started, let's create a very simple JSON theme file that alters a few of the colors in the color palette that Power BI Desktop uses for color selection.

1. Activate the View menu (or ribbon if you prefer).

2. Click the popup for the Themes. You will see something like the options shown in Figure 2-1.

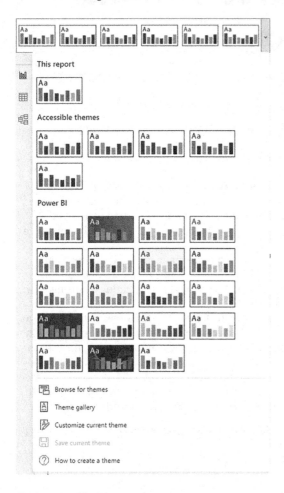

Figure 2-1. *The Themes popup*

3. Click Customize current theme. The dialog shown in Figure 2-2 will appear.

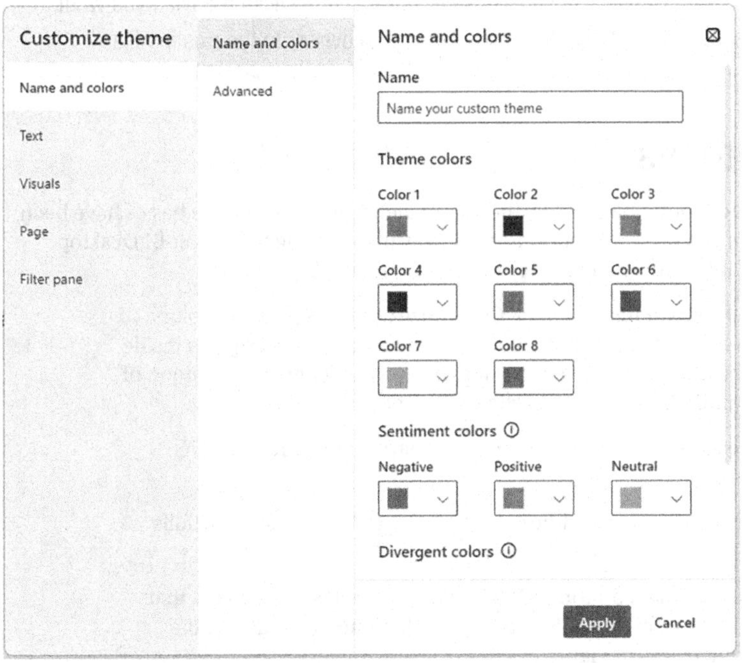

Figure 2-2. *The Customize theme dialog*

4. Select different colors for the first few Theme colors in the dialog (these correspond to the colors in the standard Power BI color palette, from left to right starting at the third column). It is worth noting that you cannot alter the first two (monochrome) colors in the color palette.

5. Click Apply. The selected colors will be applied to the current dashboard.

6. Click View ➤ Themes ➤ Save current theme. The Save As dialog will appear.

7. Browse to a suitable directory and enter a name for the theme file. This file *must* have the extension *.json*.

8. Click Save.

That is it! You now have a fully functioning JSON theme file that you can apply to other dashboards using the approach outlined in Chapter 1. You can see the effect of the theme simply by displaying the Color popup for any element in a dashboard where this theme file has been applied. You will see that the top row of the available colors has changed to display the colors that you selected in step 4 earlier (well, for all but the first two colors which will always remain white and black).

The choices that you just made are, fortunately, not set in stone. So you can alter the theme settings at any time by repeating the process that you just applied and selecting any of the available options and resaving the theme file.

___

■ **Note** When you resave a theme file, Power BI Desktop does **not** remember the name that you last used and will default to *Theme.json*. Remember to select the previous file name before clicking Save if you want to update the existing file. You will, of course, have to confirm that you wish to overwrite the existing file.

___

Name and Color Settings

A customized theme can define much more than simply the color palette. So, now that the basics have been explained, it is time to review all the customizable options that are available using the Power BI Desktop theme interface. These are all available in the dialog that you saw in Figure 2-2.

- *Name*: This field that appears at the top of the Customize theme/Name and colors tab (although it is not visible in Figure 2-2) allows you to add a text description inside the actual JSON file. This can be useful for tracking the evolution of your definition of themes – or simply reminding you what a particular theme is used for.

- *Sentiment colors*: These are used in KPI and waterfall charts to indicate positive, negative, or neutral results.

- *Divergent colors*: These are used in conditional formatting to show when data falls inside a range.

- *Advanced* (colors): The advanced options allow you to set a series of six colors that will be applied to a wide range of Power BI dashboard elements. You can see the Customize theme/Advanced tab in Figure 2-3.

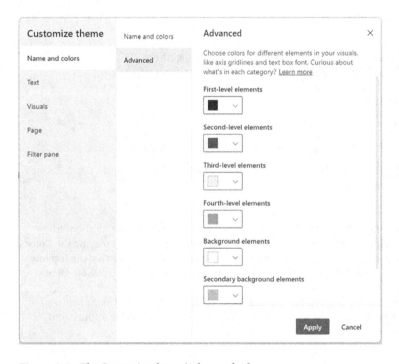

Figure 2-3. *The Customize theme/Advanced tab*

As the names of each of these color elements are somewhat abstract, here is the full list of the elements that each color setting will modify:

- *First-level elements*: Labels background color (when outside data points), trend line color, textbox default color, table and matrix values and totals font colors, data bars axis color, card data labels, gauge callout value color, KPI goal color, KPI text color, slicer item color (when in focus mode), slicer dropdown item font color, slicer numeric input font color, slicer header font color, scatter chart ratio line color, line chart forecast line color, map leader line color, filter pane and card text color

- *Second-level elements*: "Light" secondary text classes, label colors, legend label color, axis label color, table and matrix header font color, gauge target and target leader line color, KPI trend axis color, slicer slider color, slicer item font color, slicer outline color, line chart hover color, multi-row card title color, ribbon chart stroke color, shape map border color, button text font color, button icon line color, button outline color

- *Third-level elements*: Axis gridline color, table and matrix grid color, slicer header background color (when in focus mode), multi-row card outline color, shape fill color, gauge arc background color, applied filter card background color, disabled button outline color

- *Fourth-level elements*: Legend dimmed color, card category label color, multi-row card category labels color, multi-row card bar color, funnel chart conversion rate stroke color, disabled button text font color, disabled button icon line color

- *Background elements*: Labels background color (when inside data points), slicer dropdown items background color, donut chart stroke color, treemap stroke color, combo chart background color, button fill color, filter pane and available filter card background color

- *Secondary background elements*: Table and matrix grid outline color, shape map default color, ribbon chart ribbon fill color (when match series option is turned off), when background is not set to FFFFFF, the disabled button fill color, and the disabled button outline color

To give you some idea of how these formats are applied, take a look at Figure 2-4. Please note that this dashboard is not a complete reference – it just aims to show you how the principle of general color settings works.

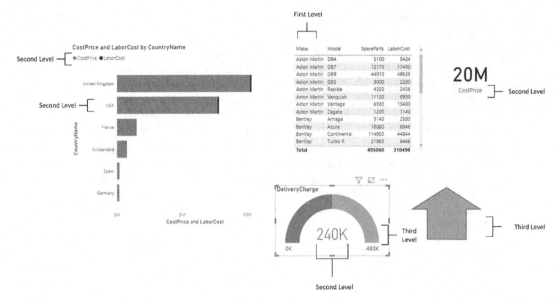

Figure 2-4. *Anatomy of high-level color customization*

Text Settings

As you might expect, there is also a set of text formats that you can define as part of a custom theme. Again, to make defining the elements easier, they are subset into a number (four in this case) of distinct elements. These subelements are

- *General*: Formats table and matrix column headers, matrix row headers, table and matrix grid, table and matrix values

- *Title*: Formats category axis title, value axis title, multi-row card title, slicer header

- *Cards and KPIs*: Formats card data labels, KPI indicators

- *Tab headers*: Formats key influencers headers

Each of the tables corresponding to one of these subelements allows you to set the

- Font

- Font size

- Font color

Let's look at each of these subelements individually.

The Customize theme/Text/General tab looks like Figure 2-5.

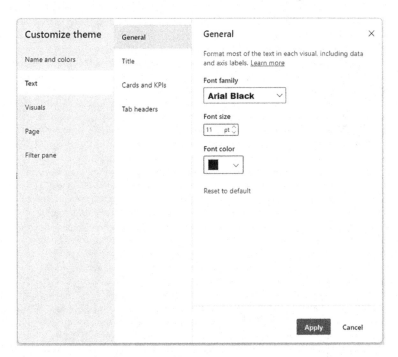

Figure 2-5. *The Customize theme/Text/General tab*

■ **Note** As you will see in subsequent chapters, this setting corresponds to the *label* element in JSON.

The Customize theme/Text/Title tab looks like Figure 2-6.

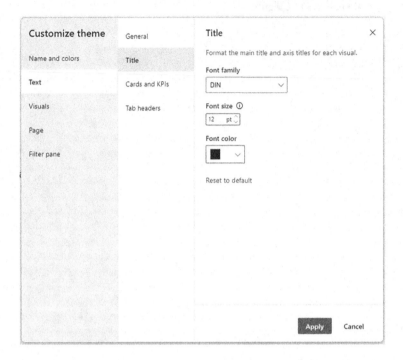

Figure 2-6. *The Customize theme/Text/Title tab*

▨ **Note** As you will see in subsequent chapters, this setting corresponds to the *title* element in a JSON theme.

The Customize theme/Text/Cards and KPIs tab looks like Figure 2-7.

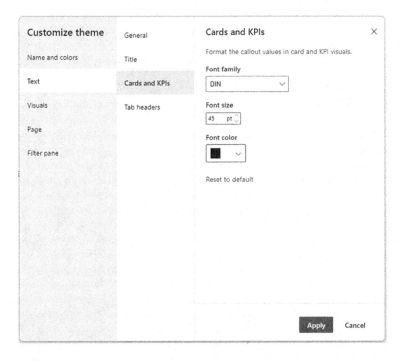

Figure 2-7. *The Customize theme/Text/Cards and KPIs tab*

■ **Note**   As you will see in subsequent chapters, this setting corresponds to the *callout* element in JSON.

The Customize theme/Text/Tab headers tab looks like Figure 2-8.

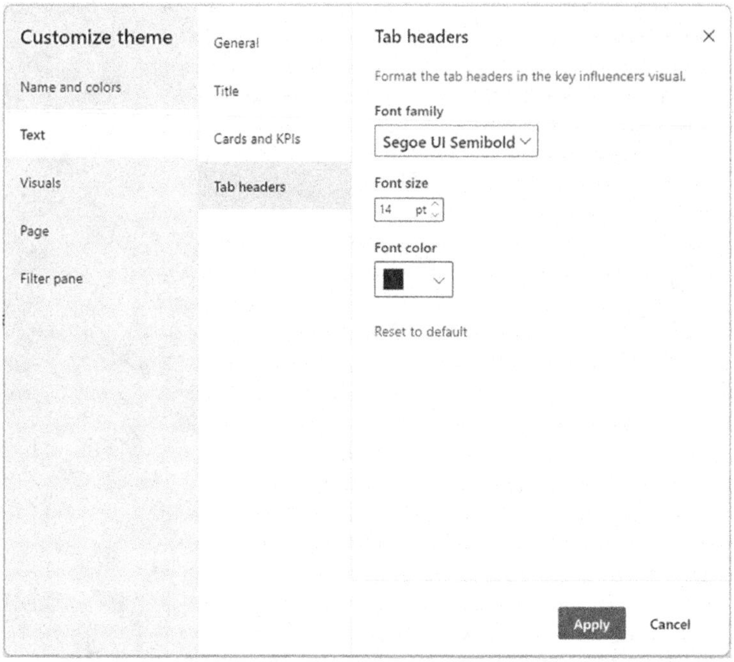

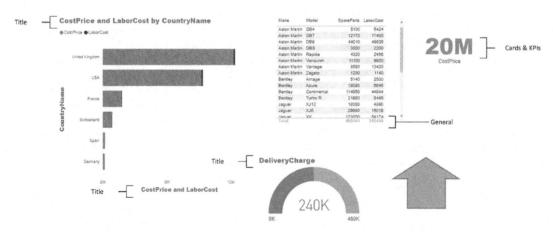

Figure 2-8. *The Customize theme/Text/Tab headers tab*

■ **Note**   As you will see in subsequent chapters, this setting corresponds to the *header* element in JSON.

To give you some idea of how these text formats are applied, take a look at Figure 2-9. Please note that this drawing is not a complete reference – once again, the idea is simply to show you how the principle of general color settings works.

Figure 2-9. *Anatomy of high-level text customization*

■ **Note**   You can see from this image that the colors defined in the name and colors section override colors set in the text section.

Visual Settings

All visuals in Power BI contain a core set of elements whose format you can define globally. This ensures a standard look and feel for all the built-in visuals that you create in a dashboard. The elements that you can define are

- *Background*: The background color and transparency of a visual

- *Border*: The visual's border attributes

- *Header*: A visual header (including its visibility)

- *Tooltip*: The colors of the popup tooltip elements

Each of these elements corresponds very closely to the standard cards (also called elements or sections) that you are used to dealing with when formatting visuals in the format pane of Power BI Desktop. They show the same attributes that you have probably tweaked many times, but presented slightly differently.

Let's take a detailed look at the way that these settings can be customized by looking at the four tabs in the Visuals section of the Customize theme dialog.

The Customize theme/Visuals/Background tab looks like Figure 2-10. This tab lets you set the color of the background for all visuals as well as the level of transparency.

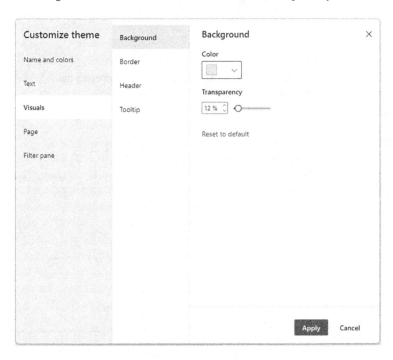

Figure 2-10.   *The Customize theme/Visuals/Background tab*

The Customize theme/Visuals/Border tab looks like Figure 2-11. As you might expect, this tab lets you specify whether a border is required, and if it is, what color the border should be. You can also define the border as having rounded corners by setting a radius figure.

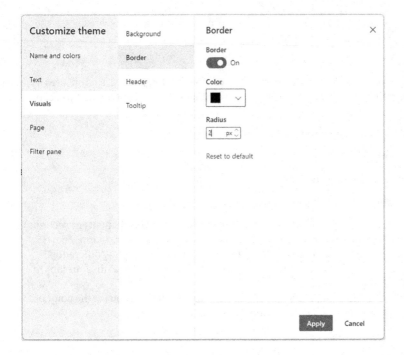

Figure 2-11. *The Customize theme/Visuals/Border tab*

The Customize theme/Visuals/Header tab (shown in Figure 2-12) allows you to set the essential attributes of all visual headers. Specifically, you can define

- The header background color
- The header border color
- The header transparency
- The color of the icons in the header

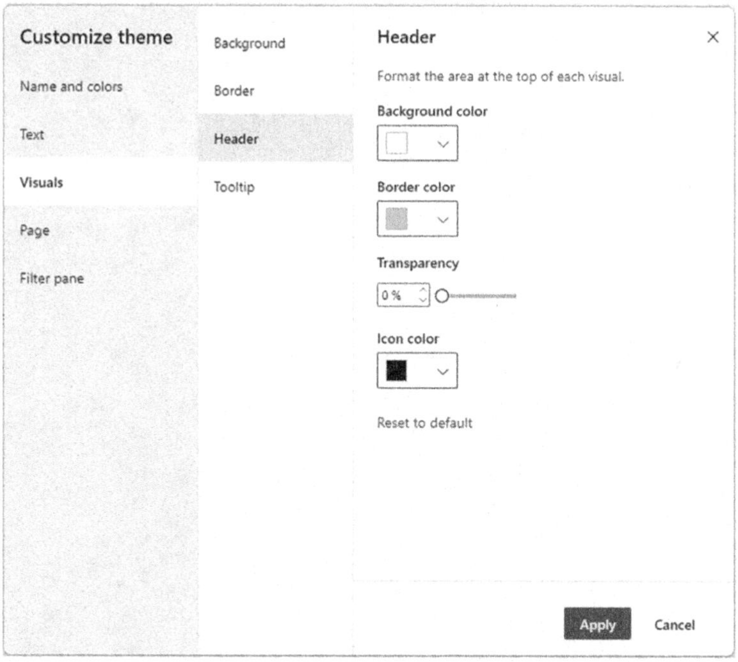

Figure 2-12. *The Customize theme/Visuals/Header tab*

The Customize theme/Visuals/Tooltip tab enables you to define the background, label, and value font colors for all popup tooltips.

It looks like Figure 2-13.

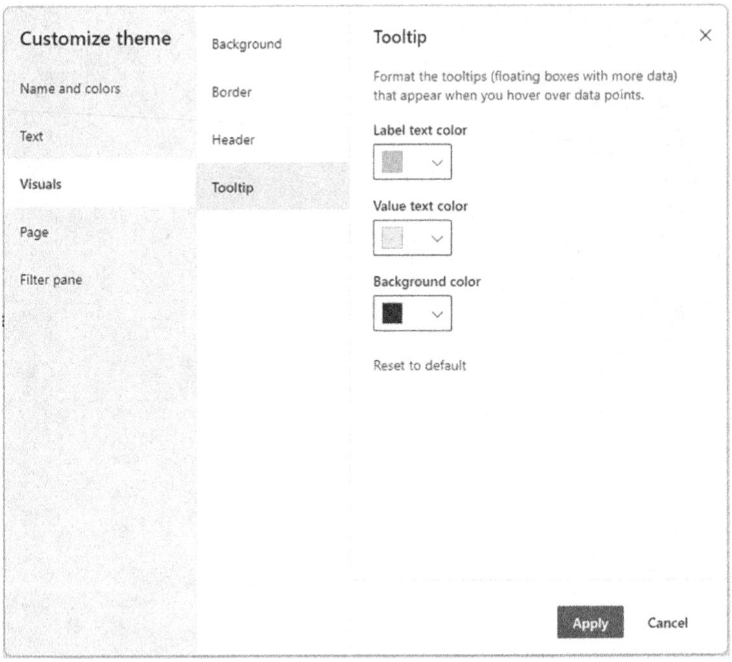

Figure 2-13. *The Customize theme/Visuals/Tooltip tab*

Page Settings

There are two settings that you can modify as far as defining the page formatting when creating a JSON theme file with Power BI Desktop is concerned. These are

- The wallpaper (the area outside the page itself)

- The page background (the color of the page underneath any visuals)

These settings are accessed using the Customize theme/Page dialog.

The Customize theme/Page/Wallpaper tab looks like Figure 2-14.

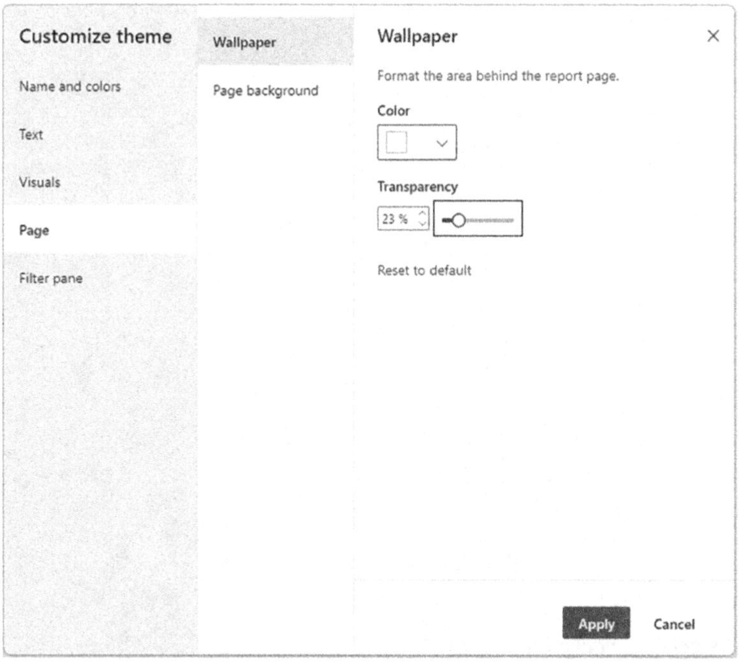

Figure 2-14. *The Customize theme/Page/Wallpaper tab*

As you can see in Figure 2-15 the Customize theme/Page/Page background tab, you can also set the color and transparency of the page background.

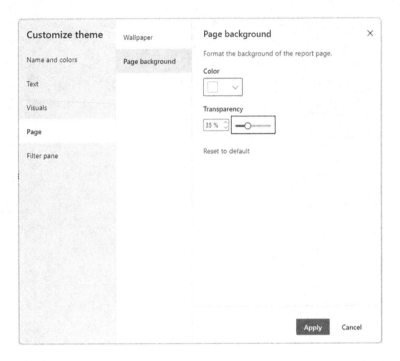

Figure 2-15. *The Customize theme/Page/Page background tab*

Filter Pane Settings

The final aspect of customized themes that you can define concerns the filter pane. Fairly logically, this is broken down into three areas:

- The filter pane itself
- Available filter cards
- Applied filter cards

The first pane that you can use is the Customize theme/Filter pane/Filter pane tab. This contains options to define

- Background color
- Transparency
- Font and icon color
- Title font size
- Header font size
- Checkbox and Apply color

You can see this in Figure 2-16.

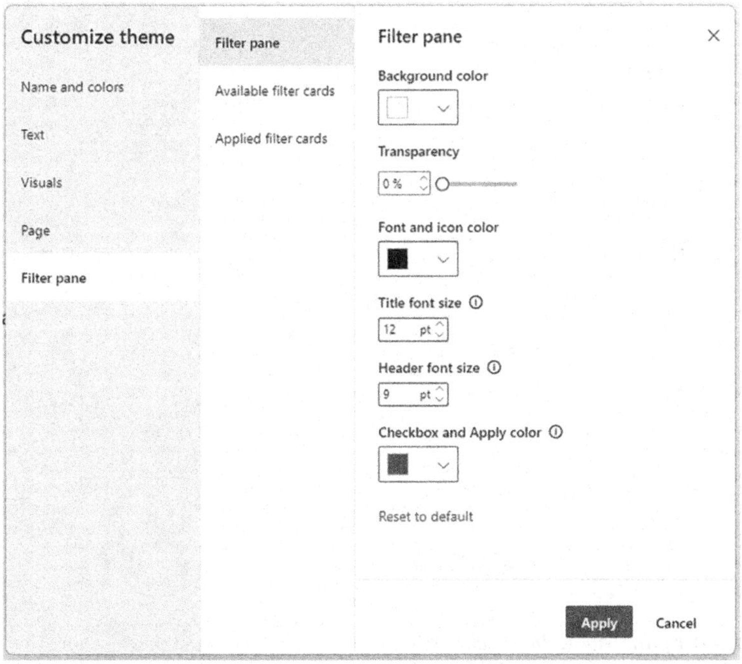

Figure 2-16. *The Customize theme/Filter pane/Filter pane tab*

The second pane (shown in Figure 2-17) that you can use is the Customize theme/Filter pane/Available filter cards tab. This contains options to define (for the filter cards inside the filter pane)

- Background color
- Transparency
- Font and icon color
- Font size

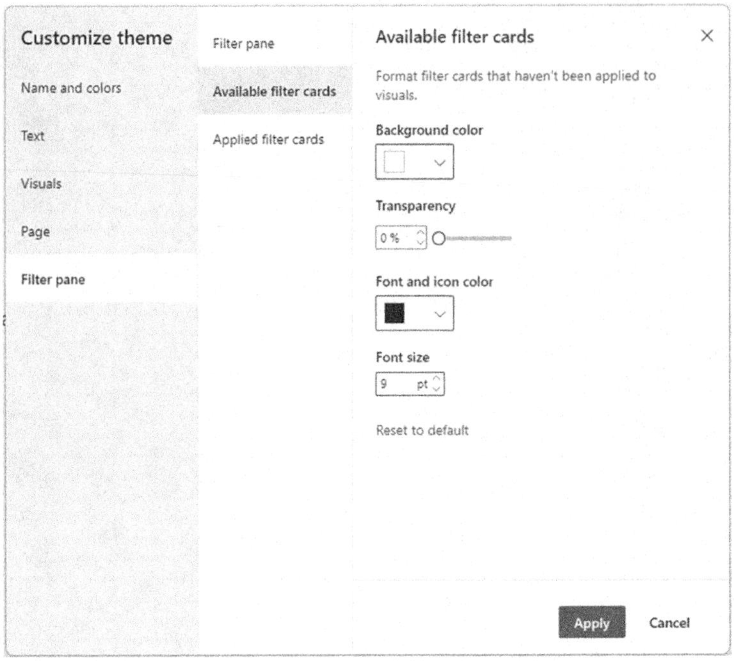

Figure 2-17. *The Customize theme/Filter pane/Available filter cards tab*

The final pane in this dialog that you can use is the Customize theme/Filter pane/Applied filter cards tab. This sets the formatting for any active filters in the filter pane. This pane, which you can see in Figure 2-18, contains options to define filter cards when a filter is set. They are

- Background color

- Transparency

- Font and icon color

- Font size

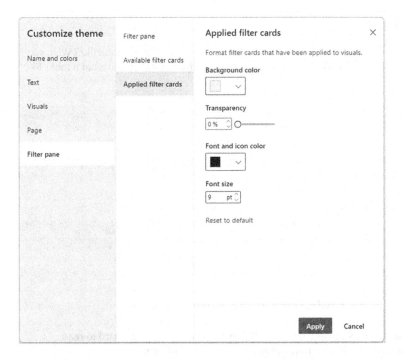

Figure 2-18. *The Customize theme/Filter pane/Applied filter cards tab*

■ **Note** Once you have set all the options that you want, remember to click the Apply button and then select View ➤ Themes ➤ Save current theme to save the JSON file that contains your settings.

Overall Comments on Interactive Theme File Creation

Now that you have seen how to create a theme file using Power BI Desktop, it is a good idea to push back and consider what you have achieved:

- You defined a high-level theme file without writing any code.

- You did this completely interactively.

In practice, there is one major takeaway that is worth remembering as far as this process is concerned:

- The theme file that you create using Power BI Desktop is less complete – and *considerably less granular* – than JSON files that you craft yourself. You can, however, extend this file using the techniques that you will learn in the remaining chapters of this book.

About the Theme File

So you have created a customized theme file in JSON format. But what does this actually look like? Well, an example file is available as the sample file PBIDesktopCustomizedTheme.json. You can see what it looks like in Figure 2-19.

Figure 2-19. *A JSON file as created by Power BI Desktop*

I am sure you will agree that the JSON created by Power BI Desktop is very ugly and hard to read. However, there are ways of making the JSON look much prettier and more readable.

To make the JSON from a generated theme file easier to read, I simply suggest that you use a search engine and look for something along the lines of "JSON prettify." You should find numerous sites that can take JSON like this and output a much more attractive version in a couple of clicks.

In a few seconds, the resulting JSON can look something like that shown in the following code snippet:

```
{
    "name": "Initial Customized Theme",
    "textClasses": {
        "label": {
            "color": "#666563",
            "fontFace": "Arial",
            "fontSize": 9
        },
        "callout": {
            "color": "#5E5C5B",
            "fontFace": "Arial",
            "fontSize": 30
        },
        "title": {
            "color": "#B8B6B4",
            "fontFace": "Arial",
            "fontSize": 11
        },
        "header": {
            "color": "#CAC5C1",
            "fontFace": "'Arial Black'",
            "fontSize": 10
        }
```

```
    },
    "dataColors": [
        "#063764",
        "#ADB6F6",
        "#E6364E",
        "#7A3F00",
        "#E044A7",
        "#6BC24D",
        "#B1D900",
        "#44CCD6"
    ],
    "foreground": "#060606",
    "foregroundNeutralSecondary": "#353534",
    "backgroundLight": "#918F8D",
    "foregroundNeutralTertiary": "#D9D6D2",
    "backgroundNeutral": "#F8EFE7",
    "bad": "#E4081F",
    "neutral": "#2BD900",
    "good": "#191CAB",
    "minimum": "#DEFDFF",
    "center": "#04D900",
    "maximum": "#2C10FF",
    "visualStyles": {
        "*": {
            "*": {
                "background": [
                    {
                        "color": {
                            "solid": {
                                "color": "#EBE5E5"
                            }
                        }
                    }
                ],
                "border": [
                    {
                        "show": true
                    }
                ],
                "visualHeader": [
                    {
                        "background": {
                            "solid": {
                                "color": "#CEC9C9"
                            }
                        },
                        "foreground": {
                            "solid": {
                                "color": "#F3EFEC"
                            }
                        },
```

```
                        "border": {
                            "solid": {
                                "color": "#0E0E0E"
                            }
                        }
                    }
                ],
                "visualTooltip": [
                    {
                        "titleFontColor": {
                            "solid": {
                                "color": "#2D2C2C"
                            }
                        },
                        "valueFontColor": {
                            "solid": {
                                "color": "#FFFFFF"
                            }
                        },
                        "background": {
                            "solid": {
                                "color": "#A29F9F"
                            }
                        }
                    }
                ],
                "outspacePane": [
                    {
                        "backgroundColor": {
                            "solid": {
                                "color": "#EE1111"
                            }
                        },
                        "foregroundColor": {
                            "solid": {
                                "color": "#222224"
                            }
                        },
                        "titleSize": 10,
                        "headerSize": 11,
                        "checkboxAndApplyColor": {
                            "solid": {
                                "color": "#1023F2"
                            }
                        }
                    }
                ],
                "filterCard": [
                    {
                        "$id": "Applied",
                        "foregroundColor": {
```

```
                        "solid": {
                            "color": "#F4820F"
                        }
                    },
                    "textSize": 10
                },
                {

                    "$id": "Available",
                    "backgroundColor": {
                        "solid": {
                            "color": "#E5E3E3"
                        }
                    },
                    "textSize": 10,
                    "transparency": 0
                }
            ]
        }
    },
    "page": {
        "*": {
            "background": [
                {
                    "color": {
                        "solid": {
                            "color": "#FAF5F5"
                        }
                    },
                    "transparency": 100
                }
            ],
            "outspace": [
                {
                    "color": {
                        "solid": {
                            "color": "#FBF8F8"
                        }
                    }
                }
            ]
        }
    }
}
```

This file is available as the sample file PBIDesktopCustomizedTheme_Prettified.json.

I realize that many readers' eyes will be glazing over at the sight of a piece of code like this. However, trust me (and continue reading) and you will soon not only be understanding JSON like this but actually creating your own. The key takeaways at this stage are that the various elements that you set (colors, text definitions, etc.) are all in the JSON file, along with the actual definitions. Exactly which JSON corresponds to what formatted element is the subject of the remainder of this book.

You may well be wondering what the outcome of all the effort that went in to creating a customized theme is. So, Figure 2-20 shows a dashboard page once the theme file PBIDesktopCustomizedTheme.json has been applied (using View ➤ Themes ➤ Browse for themes, if you remember from Chapter 1). Absolutely no other specific formatting has been applied – that is, all the formatting has been sourced from the JSON file and applied across all the visuals that have been added as well as to the underlying page and the filter pane.

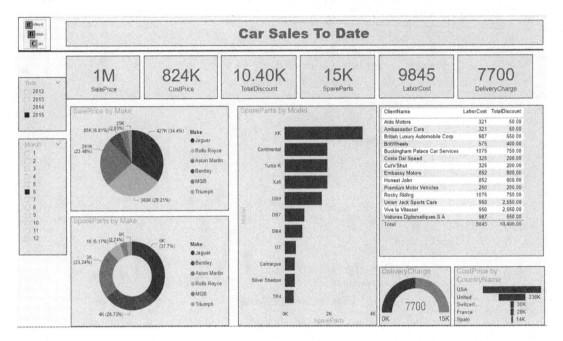

Figure 2-20. *A Power BI dashboard with a customized theme applied*

When creating your own JSON theme using the Power BI Desktop interface, you need to be aware that

- Power BI Desktop *only* creates the JSON for any elements that you modify. It will not create a theme file that defines all the elements that are capable of being set using the Customize current theme dialog. That is, the JSON output will not include any settings that you have *not* modified in the various dialogs. This makes for simpler, more concise theme files.

- Any theme file can now be applied to an existing or blank Power BI Desktop file simply by selecting View ➤ Themes ➤ Browse for themes – and navigating to the JSON file that you saved previously.

- A theme file will *only* apply formatting if an element has *not* already been formatted manually using the Format pane in Power BI Desktop. You will have to click Revert to default for each card name in each visual where you wish to remove any previously applied custom formatting or use the Reset all settings to default option.

Conclusion

In this chapter, you learned how to create and modify a JSON theme for Power BI entirely interactively using Power BI Desktop. Indeed, despite not being a completely granular definition of your formatting requirements, this approach might be all you ever need to define a perfectly usable theme file for your organization or individual requirements.

However, there is much, much more that can be done to create finely tuned themes that cover absolutely all the formatting possibilities that are present in Power BI. So, now that your appetite is whetted, it is time to move on to some progressively more detailed examples.

CHAPTER 3

■ ■ ■

High-Level Theme Definition

It is time, now, to roll up your sleeves and prepare to delve deep into the engine of Power BI themes. From now on, you will be working directly in JSON to create and modify the elements of the theme file that will control the appearance of your Power BI dashboards.

To start on your journey into Power BI themes, this chapter will take a look at the highest level of theme definitions. This involves creating generic theme settings that can apply across the board to attributes that are common to multiple visuals – or, indeed, to nearly every object in a dashboard. This could mean, for instance, setting the text attributes for column headers or axis titles – whatever the specific visual type.

There are three main advantages to this approach:

- *Firstly*, it is comparatively easy.

- *Secondly*, it applies to multiple elements – so you do not need to repeat the specification in the theme file multiple times (once for each visual type).

- *Finally*, as this approach only sets default formatting for aspects of a dashboard, you can tweak and complete the formatting of individual visuals just as you are used to doing using the format pane to set the final presentation.

What is more, setting these high-level elements may be all that you need to do to set the overall presentation of a series of dashboards. You are looking at a trade-off between in-depth (and potentially time-consuming) preparation of JSON files and delivering a quick fix that solves 80–90% of the requirements. In these circumstances, setting a high-level, more generic format can often be a perfectly valid solution. Of course, once the high-level styles have been applied, you can subsequently tweak the JSON formatting of any visuals that need further preparation and standardization – as you will discover in later chapters.

These higher-level theme definitions cover setting

- Palette colors

- Generic colors

- Generic text attributes

Power BI calls these kinds of themes "generic classes." They are well worth learning, as not only are they a great introduction to how themes work in Power BI, but they are also something that you can use in your day-to-day work with Power BI.

What Is JSON?

Let's start, however, with a (very) short sideways glance at some JSON basics. After all, you may never have met this format before. If you already know JSON, then feel free to skip the sections in this book that cover JSON. If this is not the case, then I advise you to spend the time to get a basic grasp of what JSON is and how it works as it is so fundamental to theme definition in Power BI.

© Adam Aspin 2023
A. Aspin, *Pro Power BI Theme Creation*, https://doi.org/10.1007/978-1-4842-9633-2_3

In the previous chapter, we referred to JSON copiously, without ever explaining what it was or why it is used by Power BI. JSON stands for JavaScript Object Notation. It is a lightweight data-interchange format that is solidly reliable for data exchange between systems, yet comprehensible (enough, at least) for mere humans to understand. It can be extensively structured and yet remain sufficiently pliable to store complex information. In the case of Power BI themes, this means the definition of how Power BI dashboard elements are formatted. Interestingly, when used to define Power BI theme files, it is irrelevant that the J in JSON stands for JavaScript, as in this context JSON is simply a way of defining dashboard presentation styles. So it is really the words *Object* and *Notation* that are the key terms to remember as in the context of Power BI, as JSON *describes the formatting* (notation) of *dashboard visuals* (objects).

JSON is a large subject, so this book will only introduce the aspects of JSON that are required to create theme files. Also, I intend to do this progressively, and only describe the JSON that is needed as it is required, rather than explain it all in one fell swoop. Indeed, I will be using a series of boxes to explain the necessary JSON concepts – beginning with this, first one, as follows.

JSON BASICS

JSON is based on the concept of *name/value pairs*. That is, it defines a name and what describes the name (its value).

JSON uses a *colon* to separate a name from its associated value.

JSON is based around the concept of objects – and, at the highest level, the entire theme is an object.

Objects can contain other (nested) objects.

All *objects* are enclosed by curly brackets (or braces if you prefer) **{ }**.

JSON uses *commas* to separate lists of elements.

JSON can contain *arrays*. These are lists of elements enclosed in square brackets **[ ]**.

JSON insists that you enter texts in double quotes and numbers without quotes.

Palette Colors

You are now about to create your first Power BI theme file. It will be extremely simple. Indeed, it is probably easier to load the sample file *PaletteColors.json* from the directory containing the sample files and modify it rather than to try and recreate the entire file by hand from scratch. To be practical, this is a technique that I advise you to use whenever possible when creating theme files – begin with something that works and adjust and/or extend it to suit your new requirements.

Figure 3-1 shows the contents of the sample file *PaletteColors.json*.

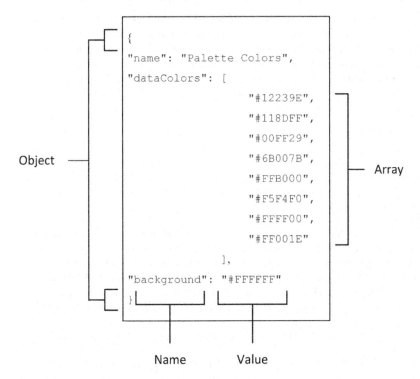

Figure 3-1. *The JSON to set the color palette*

The theme file contains three core JSON structures:

- Curly braces that enclose the entire *JSON object*.

- Name/value pairs (*name, dataColors,* and *background* are the *names* – the text or array following the colon in each case is the *value*).

- An *array* (the list of comma-separated hexadecimal colors inside square brackets). Here, the entire array is the list of values for the name/value pair.

As this is your first encounter with a theme file, I prefer to explain everything that it contains, as this will help you to understand how the JSON works.

The JSON Object

The entire JSON "object" (as it is called) must start and end with opening and closing curly braces. There should, ideally, be *no text at all* either before or after the initial opening and final closing curly braces – though if you do add extra empty lines, it will not matter as long as there is no added text at all.

There may be further nested sets of curly braces that define other objects contained inside the top-level pair of braces. Indeed, as you examine more complex JSON theme files in upcoming chapters, you will see examples of this.

Attributes Defined by the Theme File

This particular theme file defines two sets of formatting attributes (*dataColor* and *background*) and contains an added element (*name*) that describes the file (after all, files full of code are not necessarily easy to distinguish one from another). Each element in the file is defined by a structure called a name/value pair. It consists of a formatting element name (or keyword) in double quotes, separated by a colon from the value that this keyword describes. The value that is described can range from the extremely simple (a figure or a word) to a much more complex JSON structure – like the array that you can see in this theme file.

The two sets of formatting attributes that are defined are

- *dataColors*: The color palette

- *background*: The page background color

The first thing to note is that each formatting element (or keyword) is in *double quotes and followed by a colon* and finally the definition of the appropriate formatting information. This approach is also called name/value pairing (even if the value can be quite complex in itself).

The second thing is that there are *no spaces in keywords*.

The third thing to note is that there is often (but not always) a correspondence between the formatting terms that you see in the format pane in Power BI Desktop and the keyword that this element is given in the theme file. However, this correspondence can be somewhat approximate and is not always exact. Indeed, sometimes the evolution of Power BI has completely separated the keyword used in JSON from the formatting element that appears in Power BI Desktop. Consequently, I will explain the exact correspondence between the keyword used in JSON and the term used in the format pane of Power BI Desktop as this book progresses.

The fourth thing to note is that each keyword/definition pair (or name/value pair if you prefer) *ends with a comma except for the last one*.

The fifth important point is that any values that are not purely numeric *have* to be enclosed in double quotes.

The final essential point is that keywords always begin with a lowercase character – and that the spelling and case used are very unforgiving. In other words, you *must* spell the JSON keyword exactly as expected by Power BI – or the JSON attribute will *not* be applied. This means that *using the correct keyword is fundamental*. Theme files are very rigid in many ways, and as far as keywords are concerned, they are completely unforgiving.

A JSON Array

There are multiple values that define the dataColors section of the JSON file. This is because the Power BI color palette contains ten elements. The first two are invariable (white and black), whereas the other eight can be defined by the user.

These eight elements (at most – you can choose to set fewer elements and leave the others as default values) are set in the theme file as an array. This means that the list of elements is

- Enclosed in square brackets

- Comma separated – with no comma after the final element

The elements themselves can be all numbers or all text – this will depend on the kind of information that the array contains.

JSON PRESENTATION

JSON code can be displayed in an infinite number of ways. The code that you just saw could also be displayed like this:

```json
{
"name": "Intro Theme",
"dataColors": ["#12239E", "#118DFF", "#00FF29", "#6B007B", "#FFB000",
"#F5F4F0", "#FFFF00", "#FF001E"],
"background": "#FFFFFF"
}
```

or like this (among other ways):

```json
{"name": "Intro Theme", "dataColors": ["#12239E", "#118DFF", "#00FF29",
"#6B007B", "#FFB000", "#F5F4F0", "#FFFF00", "#FF001E"], "background":
"#FFFFFF"}
```

While more condensed JSON code may take up less space on the page, I prefer to make the code as readable as possible by including lots of line returns, indentations, and white space. This makes it easier to understand the correlation between the theme file and the formatting elements that are visible in the Format pane – and that the JSON file maps to.

So what have you created in this small theme file? Quite simply, you have defined the base color palette that will appear every time you want to apply colors in a Power BI dashboard. You can see this in Figure 3-2, where the hexadecimal color definitions in the *dataColors* array in the theme file define the color palette elements.

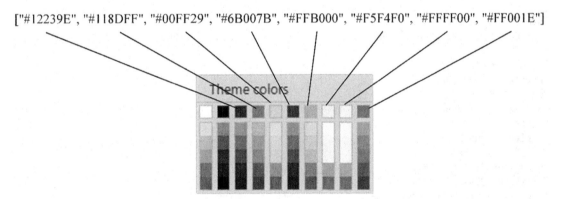

Figure 3-2. Mapping the color palette in a theme file

Colors in a JSON Theme File

The codes – each beginning with a pound/hash sign – in the attributes for the *dataColors* element are the hexadecimal values that correspond to a specific color. If you have a background in web design, then this is probably second nature. If not, then you need to find a way of finding exactly which hexadecimal value corresponds to each of the colors that you wish to use in Power BI.

Fortunately, there are literally hundreds of websites that provide guidance and tools on the use of color. So a few minutes searching for "Hex colors" in your favorite search engine should point you to some useful resources.

It is also worth knowing that the Power BI Desktop colors popup displays the hexadecimal reference of the colors that you have attributed to dashboard elements. Simply hover the pointer over a selected color in the color picker to see its hexadecimal value.

▓ **Note**   Always enclose the hexadecimal color value in double quotes and precede it with the hash/pound (#) symbol.

Generic Colors (Color Classes)

Now that you have seen an elementary JSON theme file, it is time to extend this by adding some further generic formatting elements. These will add the sentiment colors and advanced colors that were added interactively using the Customize current theme function in the previous chapter, as well as another element – *tableAccent* – that cannot be added using the Power BI Desktop interface.

You can find this file as StructuralColors.JSON in the sample files directory. The code that has been added - shown below - is in boldface.

```
{
"name": "Structural Colors",
"dataColors": ["#171796", "#E34B20", "#F4B7A6", "#465437", "#E2EB00", "#615E55", "#00AC98",
"#73268c"],
"good": "#1AAB40",
"neutral": "#D9B300",
"bad": "#D64554",
"maximum": "#118DFF",
"center": "#D9B300",
"minimum": "#DEEFFF",
"null": "#FF7F48",
"background": "#FFFFFF",
"firstLevelElements": "#949494",
"secondLevelElements": "#949494",
"thirdLevelElements": "#D5D5D5",
"fourthLevelElements": "#B7B7B7",
"secondaryBackground": "#E8E8E8",
"tableAccent": "#118DFF"
}
```

As you can see, this file extends the principles that were started with the simple file IntroTheme. json. That is

- The entire file contents are enclosed in a pair of curly braces.

- Each definition is a name/value pair.

- All definitions end with a comma except the last one.

The name/value pairs for all the color definitions (except the color palette) are simple:

- The name of the color attribute (the keyword) on the left

- The hexadecimal color reference to apply the chosen color on the right after a colon

As most of these color elements were explained in the previous chapter, I will not repeat their definition here. The only new one is

- *tableAccent*: This element overrides the table and matrix grid outline color.

It is worth noting that you can (with a simple JSON file like this one, at least) enter these elements in *any order*. The comma separation is the key point to remember – and apply rigorously. In other words, every attribute definition except the last one must end with a comma.

Interestingly, you can see that longer (and more complex) keywords are in *camelCase* – that is, they begin with a lowercase character, but each word inside the longer keyword is capitalized.

■ **Note** If you intend to create an "edgy" dark theme (with black or dark background elements), then be sure to also set the values for other structural colors and the primary text class colors to light colors – or many elements will be unreadable.

If you now apply this theme to a dashboard (one of your own or the sample dashboard SampleDashboard.pbix from the sample files), you will see how much formatting has been provided by a few lines of JSON code.

To make this clearer, take a look at Figure 3-3, where you can see how *some* of the colors that are defined in the JSON theme are applied in a Power BI Desktop dashboard.

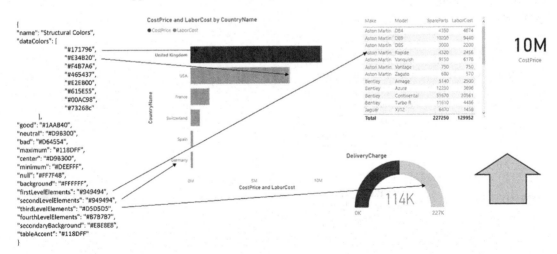

Figure 3-3. *High-level color customization*

If you wish to experiment, you can modify some of the hexadecimal color references and reapply the theme to appreciate the ease with which a range of changes can be made when you use a theme file.

Generic Text Attributes (Text Classes)

In the previous chapter, you used Power BI Desktop to create a theme file that created four text formats. These (label, callout, title, and header) were used across a range of visuals to define core text formatting. However useful, these are only some of the generic text settings that you can define using a theme file. There are another eight that you can add manually to a theme file to add a greater range of standardized text formatting to your dashboards. This can avoid you having to format manually in Power BI a whole range of text attributes, visual by visual.

Let's take a look at a JSON theme file that has added generic text classes to the color palette and generic colors that you have seen thus far in this chapter. This file is named ColorsAndTextClasses.json and is available with the sample files. The added JSON code is in boldface in the code snippet below:

```
{
"name": "Colors and Text Classes",
"dataColors": ["#171796", "#E34B20", "#F4B7A6", "#465437", "#E2EB00", "#615E55",
"#00AC98", "#73268c"],
"background": "#FFFFFF",
"good": "#1AAB40",
"neutral": "#D9B300",
"bad": "#D64554",
"maximum": "#118DFF",
"center": "#D9B300",
"minimum": "#DEEFFF",
"null": "#FF7F48",
"firstLevelElements": "#949494",
"secondLevelElements": "#949494",
"thirdLevelElements": "#D5D5D5",
"fourthLevelElements": "#B7B7B7",
"secondaryBackground": "#E8E8E8",
"tableAccent": "#118DFF",
"textClasses": {
                "label": {
                    "fontSize": 9,
                    "fontFace": "Arial",
                    "color": "#949494"
                },
                "callout": {
                    "fontSize": 18,
                    "fontFace": "Arial",
                    "color": "#808080"
                },
                "title": {
                    "fontSize": 12,
                    "fontFace": "Arial",
                    "color": "#949494"
                },
```

```
"header": {
    "fontSize": 12,
    "fontFace": "Arial",
    "color": "#808080"
},
"largeTitle": {
    "fontSize": 14,
    "fontFace": "Arial",
    "color": "#808080"
},
"largeLabel": {
    "fontSize": 12,
    "fontFace": "Arial",
    "color": "#808080"
},
"smallLabel": {
    "fontSize": 8,
    "fontFace": "Arial",
    "color": "#949494"
},
"semiboldLabel": {
    "fontSize": 8,
    "fontFace": "Arial",
    "color": "#808080"
},
"boldLabel": {
    "fontSize": 8,
    "fontFace": "Arial",
    "color": "#808080"
},
"lightLabel": {
    "fontSize": 8,
    "fontFace": "Arial",
    "color": "#949494"
},
"largeLightLabel": {
    "fontSize": 8,
    "fontFace": "Arial",
    "color": "#949494"
},
"smallLightLabel": {
    "fontSize": 8,
    "fontFace": "Arial",
    "color": "#949494"
    }
  }
}
```

So what are you looking at exactly? Essentially, this is a set of keywords that are used across a range of visuals to define text formatting. Four of them were explained in the previous chapter, as they can be created interactively using Power BI Desktop. These are label, callout, title, and header – so I will not explain the text elements that they apply to again.

The others, however, need some explanation. The following list defines which dashboard element is set by each JSON keyword:

- *largeTitle*: Visual title

- *largeLabel*: Multi-row card data labels

- *smallLabel*: Reference line labels, slicer date range labels, slicer numeric input text style, slicer search box, key influencers influencer text

- *semiboldLabel*: Key influencers profile text

- *boldLabel*: Matrix subtotals, matrix grand totals, table totals

- *lightLabel*: Legend text, button text, category axis labels, funnel, chart data labels, funnel chart conversion rate labels, gauge target, scatter chart category label, slicer items

- *largeLightLabel*: Card category labels, gauge labels, multi-row card category labels

- *smallLightLabel*: Data labels, value axis labels

To make the correlation between elements in the JSON theme file and their application to a Power BI dashboard more comprehensible, take a look at Figure 3-4, where you can see how some of the text specifications that are defined in the JSON theme are applied in a Power BI dashboard.

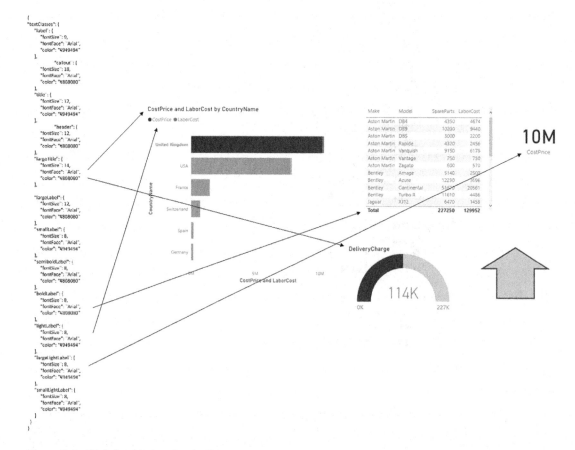

Figure 3-4. *High-level text customization*

Please note that this figure is *not* a complete representation of the application of each of the text classes. It does, however, outline how text classes work.

The JSON structure also needs some explanation, as this is the first time that you have encountered nested objects (i.e., sets of curly braces inside curly braces) in a theme file.

As you can see, the core principle is still that of a key/value pair – that is, a keyword and a definition. However, the definition is a little more complex this time, as it is not merely a single element after the colon that follows the keyword but an entire object.

You can see this schematically in Figure 3-5.

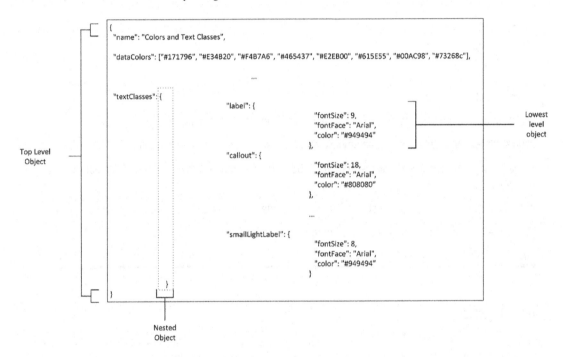

Figure 3-5. *Nested JSON objects*

The points to note are

- This time, each definition is a separate, nested object that begins and ends with *curly braces.*

- Inside each object are a series of key/value pairs that are *comma separated* (and without a comma after the final key/value pair).

- The key/value pair that makes up a text definition is, as is the case for simpler key/value pairs, *comma separated from other key/value pairs* at the "outer" level.

- *Numbers are not enclosed in double quotes.* As Power BI always expects that fonts will be defined in points, you do not need to specify anything other than the font size number (no need for "pt" or anything like that).

- Any *fonts* that you reference must be *spelled exactly as Windows expects* them and must, of course, exist in the system that is hosting the Power BI dashboard.

JSON DATA TYPES

Any values (the part after the colon in a name/value pair) must be a recognizable JSON data type.

JSON has a limited available set of data types:

- Number
- String
- Boolean
- Objects
- Array
- Null

Numbers can be integers, floating point, or decimals. They should *not* be enclosed in quotes.

Strings (or text or characters) can be any text and *must* be enclosed in double quotes.

Boolean simply means true or false – and has to be represented by the lowercase word *true* or *false* (not yes/no or 0/1, for instance).

A JSON element can be a nested object – that itself contains further elements such as key/value pairs.

Arrays are a series of elements (strings, numbers, objects, or even other arrays).

When you want to define a completely empty value, it is defined as *null* (in lowercase – without quotes).

Conclusion

This chapter started you on the path to creating and modifying your own custom JSON theme files. You saw how to add color definition and generic text attributes as JSON in a theme file.

Hopefully, the progression from the short file that this chapter began with to the final complete theme file has convinced you that crafting JSON theme files need not be complex or difficult. As the structure of a theme file is based on a clear logic, a little practice should enable you to move on to more advanced formatting definitions. Simple themes with default visual styles like those you saw in this chapter can subsequently be extended through adding specifications for standard elements across a range of types of visuals. This is the subject of the next chapter.

CHAPTER 4

■ ■ ■

Default Visual Styles

The next step in defining core Power BI dashboard formatting is to configure a set of high-level presentation elements that can cover the majority of the formatting requirements of dashboard visuals. This will ensure that much (although possibly not all) of the essential formatting of a dashboard is carried out simply through applying a JSON theme file. This is a practical application of the so-called 80/20 rule – where 80% of the formatting work can be done generically, leaving the rest to be formatted manually.

In practice, it can often be extremely effective to define a handful of "default" stylistic elements that apply to a range of visuals rather than defining the exact specification for 30 or so separate visuals. This is made possible by the fact that Power BI has ensured that any aspect of a visual that is common to other visuals (think borders or backgrounds, for instance) uses the *same* formatting approach in the Power BI Desktop format pane *whatever the visual* that you are tweaking – and this commonality is extended to the JSON theme files.

What this approach means, practically, is that you can specify a format element *once only* in a theme file and nonetheless see it applied to *any visual* that supports it. For instance, you may want to define borders and titles for all visuals and then specify a handful of common elements (such as X and Y axis definitions) for charts as well as core table elements such as formatting totals and values. Moreover, you do not even have to specify the entire gamut of formatting that is available for a visual in Power BI. That is, if you only want to define that a title is required (without defining any other characteristics of the title), then you can limit the specification to this single attribute.

All of this is possible through a handcrafted theme file where you define visual characteristics *without* specifying which visual type the formatting must apply to. The JSON format will be applied – instantly and automatically – to *any visual* that contains the attributes that you define. There are three great things about this approach:

- It simplifies theme preparation, as you only define an element *once* (as opposed to defining it individually for each type of specific visual).

- You can make the definition of the formatting as *detailed or high level* as you wish and then tweak individual format options in Power BI Desktop to add completeness.

- You can add default formatting "in case." This is because any formatting that you add that is *not* used by a visual type (such as defining certain detailed elements of a certain chart type) will not prevent other formats being applied where they *are* supported by a different chart type.

What is more, you can subsequently overwrite or extend these common attributes at the level of each individual visual if you so wish. These techniques are described in Chapters 5 through 12.

In this chapter, we will, progressively, see how to

- Define visual elements common to most types of visual

- Define elements common to most chart types

- Define table elements

© Adam Aspin 2023
A. Aspin, *Pro Power BI Theme Creation*, https://doi.org/10.1007/978-1-4842-9633-2_4

Creating default visual styles means understanding the JSON formatting of each individual element in Power BI. Clearly, this can require a full understanding of the intricacies of JSON theme file creation. So, what you will be learning in this chapter is the *principle* of how to select core aspects of JSON formatting code in order to apply them at a more abstract level so that the JSON can shape the presentation of multiple visuals – *without having to define each one individually*. The in-depth appreciation of all the details specific to each individual visual will come later in the book.

Common Visual Elements

There are a series of elements that are common to most – if not all – visuals in Power BI dashboards. These elements are displayed in the General tab of the Format pane and consist of

- Properties (Padding, Lock Aspect, and Responsiveness)
- Title – including Subtitle, Divider, and Spacing
- Effects, consisting of Background, Visual Border, and ShadowHeader Icons (including Header Icon display and Help Tooltips)
- Tooltips

You are probably used to setting these – time and time again, if the truth be told – using the Power BI format pane.

So the time has come to define these common elements using JSON in a theme. As is the case with most of the approaches to formatting Power BI dashboards in this book, it is best to take a look at the JSON code first and explain how it works once you have seen the code. So here is a code snippet (with new elements in boldface) that adds some core visual default formatting to elements that are common to most visuals.

This code snippet hides some of the JSON that you have seen previously to save space on the page. The complete code is in the file CommonElements.json.

You have to be aware at the outset that there is not a one-to-one mapping between the JSON keywords used in a theme file and the actual sections of the Power BI General format pane. The formatting elements are defined by the following JSON theme keywords:

- The *title* keyword corresponds to the Title subsection of the Title section.
- The *subTitle* keyword corresponds to the Subtitle subsection of the Title section.
- The *divider* keyword corresponds to the Divider subsection of the Title section.
- The *spacing* keyword corresponds to the Spacing subsection of the Title section.
- The *lockAspect* keyword corresponds to the Lock aspect subsection of the Properties section.
- The *background* keyword corresponds to the Background subsection of the Effects section.
- The *border* keyword corresponds to the Border subsection of the Effects section.
- The *dropShadow* keyword corresponds to the Shadow subsection of the Effects section.
- The *visualTooltip* keyword corresponds to the Tooltips section.
- The *visualHeader* keyword corresponds to the Icons subsection of the Header Icons section.
- The *visualHeaderTooltip* keyword corresponds to the Help Tooltip subsection of the Header Icons section.

```
{
"name": "Common elements",
"dataColors": ["#12239E", "#118DFF", "#00FF29", "#6B007B", "#FFB000", "#F5F4F0", "#FFFF00",
"#FF001E"],
    --- Code not shown for color and text classes ---
"visualStyles":
  {"*":
   {"*":
    {
   "title": [{
       "show": true,
       "text": "Visual Title",
          "heading": "Heading2",
       "fontColor": {"solid": {"color": "#ffffff"}},
       "background": {"solid": {"color": "#12239E"}},
       "alignment": "Center",
       "fontSize": 8,
       "fontFamily": "Arial",
          "bold": true,
          "italic": false,
          "underline": false,
          "titleWrap": true
       }],
    "subTitle":     [{
       "show": true,
       "text": "Chart Subitle",
       "heading": "Heading3",
       "fontColor": {"solid": {"color": "#ffffff"}},
       "background": {"solid": {"color": "#12239E"}},
       "alignment": "left",
       "fontSize": 9,
       "fontFamily": "Arial",
       "bold": true,
       "italic": false,
       "underline": false,
       "titleWrap": true
             }],
    "divider":     [{
       "show": true,
       "color": {"solid": {"color": "#615E55"}},
       "style": "solid",
       "width": 2,
       "ignorePadding": true
             }],
    "spacing": [{
       "customizeSpacing": true,
       "spaceBelowTitle": 3,
       "spaceBelowSubTitle": 3,
       "spaceBelowTitleArea": 2
             }],
```

47

```
"lockAspect": [{
    "show": false
      }],
    "background": [{
       "show": true,
       "color": {"solid": {"color": "#F5F4F0"}},
       "transparency": 0
           }],
"border":    [{
    "show": true,
    "color": {"solid": {"color": "#F5F4F0"}}
      }],
"dropShadow":    [{
    "show": true,
    "color": {"solid": {"color": "#E8E8E8"}},
    "position":"Outer",
    "preset":"TopLeft"
      }],
"visualHeaderTooltip":[{
    "type": "Default",
    "titleFontColor":{"solid":{"color":"#E34820"}},
    "text": "Add a tooltip text for the visual",
    "fontSize": 11,
    "fontFamily": "Arial Black",
    "background":{"solid":{"color":"#D1CAB8"}},
    "transparency": 8,
    "bold": true,
    "italic": false,
    "underline": false
        }],
"visualHeader": [{
    "showVisualInformationButton": false,
    "showVisualWarningButton": false,
    "showVisualErrorButton": false,
    "showDrillRoleSelector": true,
    "showDrillToggleButton": false,
    "showDrillUpButton": false,
    "showDrillDownLevelButton": false,
    "showDrillDownExpandButton": false,
    "showPinButton": false,
    "showFocusModeButton": false,
    "showFilterRestatementButton": false,
    "showSeeDataLayoutToggleButton": false,
    "showOptionsMenu": false,
    "showTooltipButton": true,
    "showSmartNarrativeButton": true
          }],
"visualTooltip": [{
    "show": true,
    "type": "Default",
    "fontSize": 11,
```

```
      "fontFamily": "Arial Black",
      "bold": true,
      "italic": false,
      "underline": false,
      "background": {"solid": {"color": "#ffffff"}},
      "transparency": 10,
      "titleFontColor": {"solid": {"color": "#E8Ef33"}},
      "ThemeDataColor": {"solid": {"color": "#E8E8E8"}}
            }],
   "general":  [{
      "responsive": true,
      "altText": "Enter an alternative text"
            }],
   "padding": [{
      "left":3,
      "right":3,
      "top":3,
      "bottom":3
         }]   }
  }
 }
}
```

Now for some explanations concerning this particular piece of JSON

- Each JSON attribute corresponds to an element in a specific section or subsection of the format pane in Power BI Desktop.

- The core method of defining key/value pairs is still respected. However, as you can see, each of these value settings is defined as an *array* that contains one or more objects. That is, each definition starts with a square bracket and a curly bracket and ends with a curly bracket followed by a square bracket.

- Some objects contain further nested objects – that is, inner sets of elements in curly brackets, sometimes up to three objects deep.

- There is something else that is new in this JSON code. You are referring to selections made using popup menus in Power BI Desktop to format elements such as

  - Title alignment

  - Shadow position

  - Shadow preset

- You need to be warned that the *exact spelling of each JSON keyword is required* – and this must always be enclosed in double quotes. Of course, if you don't want to guess the definition, then you can look them up in the chapter that describes the visual that you want to format.

- The good news about these definitions is that they do *not* cause the JSON file load to fail if you misspell one of them. All that will happen is that the actual attribute will *not* be formatted and will remain as the default format once the JSON theme file has been loaded.

- You can see that whereas Power BI Desktop has a button to switch elements on or off (such as titles or shadows), the corresponding JSON translates as

```
"show": true
```

or

```
"show": false
```

- You need to be aware that JSON displays Boolean values (i.e., true or false):
  - Without quotes
  - In lowercase
- It is important to respect the capitalization for the words true and false as JSON themes are *case-sensitive*. You must *not* place them in double quotes either, as they are keywords, not values.
- Finally, the actual element definitions (background, title, etc.) are, themselves, nested inside two "anonymous" elements – as shown in Figure 4-1.

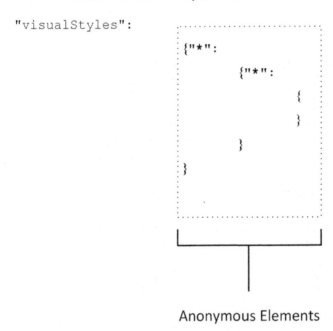

Figure 4-1. *Anonymous elements*

When defining generic elements (i.e., formatting not strictly attributed to a specific type of visual), they *have* to be nested at this lower level. The asterisks mean that the formatting is generic – not tied to a specific object. This does admittedly make tracking the nesting levels (and all the pairs of curly brackets) a little more complex. However, it is something that you get used to the more that you work with JSON theme files.

JSON Attribute Definition

If you take a closer look at the JSON, you will notice that it is really quite easy to understand, as each keyword/value pair for each section's formatting options is reasonably straightforward. What the JSON is doing is setting each attribute individually for each section. Moreover, the JSON follows the principles that you saw previously:

- On or off is indicated using "show": true or "show": false.

- Sizes are simple, unquoted numbers.

- Popup selections are in quotes – and virtually identical to the actual choice in any popup menus.

- Colors are hexadecimal values in double quotes (preceded by a hash/pound symbol).

This means that understanding the JSON that is used comes easily to a Power BI Desktop user who has already spent time formatting charts in dashboards. Just to reinforce an important point, the JSON that you see here is only a subset of the formatting options that are available for most of these keywords. In Chapters 5 through 9, you will see the full details of all currently available chart formatting options for each chart type.

JSON Keywords

The fact that theme files use keywords inevitably means that you need to know the exact terms that are required in a JSON theme file if you are going to set formatting attributes that

- Map to a section in the format pane of Power BI Desktop

- Specify an attribute

- Are chosen from a popup menu in Power BI Desktop

The general rules for guessing the definition in JSON of a selectable element in Power BI Desktop are

- Hover the pointer over an icon (such as the alignment icons) and use the popup text as the definition

- Use the popup menu text (for the shadow position, for instance) but with all spaces removed and with initial capitals

However, to save you guessing at keywords, they are all described in later chapters of this book.

The JSON Hierarchy in Theme Files

JSON lends itself to hierarchical structures – and so is based on the approach that you can see outlined in Figure 4-2.

Figure 4-2. *The JSON hierarchy in theme files*

▪ **Note** The style name is always an asterisk. If you are defining generic attributes (as is the case in this chapter), then the second asterisk replaces the visual name that you can see in later, more specific, theme files.

Elements Common to Most Chart Types

Now that you have seen how to define the formatting for shared (or common) elements, we can move on to the next level of "generic" formatting. This involves setting *some* of the formatting parameters of chart visuals.

There are several aspects of chart formatting that are shared by multiple charts. For instance, stacked bar charts, stacked column charts, clustered bar charts, clustered column charts, line charts, area charts, stacked area charts, line and clustered column charts, and line and stacked column charts all contain the following elements:

- The category (X) axis
- The values (Y) axis
- Data labels
- Plot area
- Legend

Indeed, certain of these elements are also common to a few other chart types such as ribbon charts, waterfall charts, and scatter charts (even if these chart types only contain some of the formatting options in the preceding list). There are even charts such as funnel charts, pie charts, and donut charts that use at least one of the formatting elements listed earlier.

To save you the effort of having to define these similar or identical aspects of chart formatting to each individual chart type, you can specify *standard formatting* that will be applied to a chart element *if it exists*. This means that if you have created a chart that does not have one of these elements (say a donut chart that does not have category and values axes but *does* have a legend) but have nonetheless set these attributes in a theme file, then the definitions of any nonrelevant formatting elements are simply *not applied* to the chart in question. What is more, you do *not have to define every format option* that exists in an expanded tab of the Format Pane. That is, you can only decide to set one or two formatting options for the legend out of all the available options, if you so wish. In other words, you can predefine as many aspects of chart formatting as you want without worrying about actually having to use them. Power BI Desktop does not care about whether a JSON format is used – only if it is syntactically accurate. This means that you can define as much (or as little) generic formatting as you like as long as the JSON is correct.

Here, then, is an extract from a JSON theme file that applies some (but not all) of the formatting to a series of shared chart elements. You can find the complete code in the file ColorAndTextClassesWithCoreVisualDefaultsAndKeyChartTypes.json.

The JSON to define a set of standard chart formats could look like this:

```
{
"name": "Common chart elements",
"dataColors": ["#12239E", "#118DFF", "#00FF29", "#6B007B", "#FFB000", "#F5F4F0", "#FFFF00",
"#FF001E"],
"visualStyles":
  {"*":
    {"*":
     {
  "title": [{... code not shown ...}],
  "lockAspect": [{... code not shown ...}],
  "dropShadow":   [{... code not shown ...}],
  "visualHeader": [{... code not shown ...}],
  "categoryAxis": [{
       "show": true,
       "axisScale": "Linear",
       "position": "left",
       "labelColor": {"solid": {"color": "#949494"}},
       "fontFamily": "Arial",
       "fontSize": 8,
       "showAxisTitle": true,
       "titleText": "Add an Axis Title",
       "titleFontSize": 8,
       "titleFontFamily": "Arial",
       "titleColor": {"solid": {"color": "#949494"}},
       "gridLineShow": true,
       "gridLineColor":  {"solid": {"color": "#949494"}},
       "gridLineThickness": 1,
       "gridlineStyle": "Solid"
       }],
  "valueAxis": [{
       "show": true,
       "position": "Left",
       "axisScale": "Linear",
       "labelColor": {"solid": {"color": "#949494"}},
       "fontFamily": "Arial",
       "fontSize": 8,
       "showAxisTitle": true,
       "titleText": "Add an Axis Title",
```

```
        "titleFontSize": 8,
        "titleFontFamily": "Arial",
        "titleColor": {"solid": {"color": "#949494"}},
        "gridLineShow": true,
        "gridLineColor": {"solid": {"color": "#949494"}},
        "gridLineThickness": 1,
        "gridlineStyle": "Solid",
        "axisStyle": "ShowTitleOnly"
        }],
  "labels": [{
        "show": true,
        "labelPosition": "InsideEnd",
        "color": {"solid": {"color": "#ffffff"}},
        "fontFamily": "Arial",
        "fontSize": 7,
        "enableBackground": false,
        "backgroundColor": {"solid": {"color": "#E8E8E8"}},
        "backgroundTransparency": 0
        }],
  "plotArea": [{
        "transparency": 0
        }],
  "legend": [{
        "show": true,
        "position": "Bottom",
        "showTitle": true,
        "titleText": "Add a legend title",
        "legendColor": {"solid": {"color": "#949494"}},
        "fontFamily": "Arial",
        "fontSize": 8
        }]
    }
   }
  }
}
```

In this JSON snippet – once again – the keywords correspond to sections of the Format Pane in Power BI Desktop. The sections that are referred to here are common to bar, column, line, and area charts.

- The *category (X) axis* is defined by the keyword *categoryAxis*.

- The *values (Y) axis* is defined by the keyword *valueAxis*.

- *Data labels* are defined by the keyword *labels*.

- The *plot area* is defined by the keyword *plotArea*.

- The *legend* is defined by the keyword *legend*.

Extending Generic Formatting for Specific Chart Types

The previous JSON snippet showed how to define core formatting for five chart elements. There could be cases, however, where you want to add a few elements that only apply to a select few types of chart – or even a single chart type. Fortunately, this is a simple extension to the JSON code.

Suppose, for instance, that you want to add the following two elements to the theme file:

- The line and join formatting found in line charts, area charts, stacked area charts, line and clustered column charts, and line and stacked column charts

- The central radius percentage used in donut charts

You can do this by extending the previous JSON with the following code snippet (and ensuring that there is a comma separating the two pieces of code in the JSON file). The complete JSON theme file can be found with the sample files. It is called ColorAndTextClassesWithCoreVisualDefaultsAndKeyChartTypesExtended. json. The code that is added to the snippet that you saw previously is

```json
"lineStyles": [{
    "ShadedArea": true,
    "strokeWidth": 1,
    "lineStyle": "Solid",
    "showMarker": true,
    "markerShape": "Triangle",
    "markerSize": 2,
    "markerColor": {"solid": {"color": "#949494"}}
}],
"slices": [{
    "innerRadiusRatio": 20
}]
```

This JSON will only be applied to charts that actually contain the following formatting sections in Power BI Desktop:

- *Shapes* (represented by the lineStyles keyword in JSON for line and area charts)

- *Shapes/Inner radius* (represented by the slices keyword for donut charts)

As you are starting to see, the correlation between the Power BI Desktop formatting section and the corresponding JSON keywords is not always completely intuitive. However, once the mappings are understood, creating and modifying JSON theme files becomes fairly straightforward.

Table Elements

Not all visuals in Power BI, of course, are charts. There are core visuals that are very unlike charts that you probably rely on in your dashboards such as tables, matrices, and cards. Fortunately, it is also possible to define the core formatting of these types of visuals quickly and easily.

The following code snippet adds the formatting specification for the following table (and matrix) elements:

- Grid

- Column headers

- Values

- Total

Fortunately, the JSON keywords correspond very closely to the section headers in the Power BI Desktop format pane for tables and matrices. Also, the format elements are simple key/value pairs that map closely to the actual format attributes in Power BI Desktop, so little additional explanation is necessary here.

```
"style":  [{"styleType": "Minimal"}],
"grid":    [{
    "gridVertical": false,
    "gridVerticalColor": {"solid": {"color": "#171796"}},
    "gridVerticalWeight": 1,
    "gridHorizontal": false,
    "gridHorizontalColor": {"solid": {"color": "#171796"}},
    "gridHorizontalWeight": 1,
    "textSize": 8,
    "rowPadding": 1,
    "outlineColor": {"solid": {"color": "#ffffff"}},
    "outlineWeight": 1
    }],
"columnHeaders": [{
    "fontColor": {"solid": {"color": "#ffffff"}},
    "backColor": {"solid": {"color": "#E34B20"}},
    "outline": "Frame",
    "alignment": "Left",
    "wordWrap": true
  }],
"values": [{
    "fontColorPrimary": {"solid": {"color": "#3D4E57"}},
    "backColorPrimary": {"solid": {"color": "#F5F4F0"}},
    "fontColorSecondary": {"solid": {"color": "#3D4E57"}},
    "backColorSecondary": {"solid": {"color": "#E3DFD4"}},
    "outline": "Frame",
    "outline": "Frame",
    "alignment": "Left",
    "wordWrap": true
  }],
"total": [{
    "totals": true,
    "fontColor": {"solid": {"color": "#000000"}},
    "backColor": {"solid": {"color": "#D1CAB8"}},
    "outline": "Frame"
  }]
```

Here, once again, not all the format options that are available in Power BI Desktop as far as totals, values, column headers, and grids have been applied using the JSON theme file. Once again, any format element that is not set in the JSON file is kept as the default in the Power BI Desktop file to which this theme is added.

This JSON code is found in the sample file ColorsAndTextClassesWithVisualDefaults.json.

Final Comments

If you add all the various JSON snippets that you have explored in this chapter to a single theme file, the complete JSON theme file will contain a high-level format definition of

- Color palette

- Overall color application

- Standard font specification

- Essential chart elements

- Core table elements

A file that contains all these elements assembled in a working theme is available in the sample data directory as GenericBaseFormatting.json.

You can see the entire file as follows. I realize that this is a large piece of code (some 230-odd lines) – but it is worth getting a complete overview of all that you have created and how it is assembled into a coherent theme:

```
{
"name": "Generic Base Formatting",
"dataColors": ["#171796", "#E34B20", "#F4B7A6", "#465437", "#E2EB00", "#615E55", "#00AC98",
"#73268c"],
"good": "#1AAB40",
"neutral": "#D9B300",
"bad": "#D64554",
"maximum": "#118DFF",
"center": "#D9B300",
"minimum": "#DEEFFF",
"null": "#FF7F48",
"background": "#FFFFFF",
"firstLevelElements": "#949494",
"secondLevelElements": "#949494",
"thirdLevelElements": "#D5D5D5",
"fourthLevelElements": "#B7B7B7",
"secondaryBackground": "#E8E8E8",
"tableAccent": "#118DFF",
"textClasses": {
    "label": {
        "fontSize": 9,
        "fontFace": "Arial",
        "color": "#949494"
    },
      "callout": {
        "fontSize": 18,
        "fontFace": "Arial",
        "color": "#808080"
    },
    "title": {
        "fontSize": 12,
        "fontFace": "Arial",
        "color": "#949494"
    },
```

```
        "header": {
            "fontSize": 12,
            "fontFace": "Arial",
            "color": "#808080"
        },
        "largeTitle": {
            "fontSize": 14,
            "fontFace": "Arial",
            "color": "#808080"
        },
        "largeLabel": {
            "fontSize": 12,
            "fontFace": "Arial",
            "color": "#808080"
        },
        "smallLabel": {
            "fontSize": 8,
            "fontFace": "Arial",
            "color": "#949494"
        },
        "semiboldLabel": {
            "fontSize": 8,
            "fontFace": "Arial",
            "color": "#808080"
        },
        "boldLabel": {
            "fontSize": 8,
            "fontFace": "Arial",
            "color": "#808080"
        },
        "lightLabel": {
            "fontSize": 8,
            "fontFace": "Arial",
            "color": "#949494"
        },
        "largeLightLabel": {
            "fontSize": 8,
            "fontFace": "Arial",
            "color": "#949494"
        },
        "smallLightLabel": {
            "fontSize": 8,
            "fontFace": "Arial",
            "color": "#949494"
        }
    },
"visualStyles":
    {"*":
        {"*":
            {
            "background":   [{
```

```
        "color": {"solid": {"color": "#F5F4F0"}},
        "transparency": 0
                    }],
        "visualTooltip":[{
                "titleFontColor":{"solid":{"color":"#D7CACA"}},
                "valueFontColor":{"solid":{"color":"#FFFFFF"}},
                "background":{"solid":{"color":"#938888"}}}],
                "background":[{"color":{"solid":{"color":"#7A0A0A"}}
                }],
    "title":    [{
            "show": true,
            "text": "Visual Title",
            "fontColor": {"solid": {"color": "#ffffff"}},
            "background": {"solid": {"color": "#171796"}},
            "alignment": "Center",
            "fontSize": 8,
            "fontFamily": "Arial"
                }],
    "lockAspect": [{
    "show": false
    }],
"border": [{
 "show": true,
 "color": {"solid": {"color": "#171796"}}
 }],
"dropShadow": [{
 "show": true,
 "color": {"solid": {"color": "#D7CACA"}},
 "position":"Outer",
 "preset":"TopLeft"
 }],
"visualHeader": [{
 "show": true,
 "background": {"solid": {"color": "#ffffff"}},
 "border": {"solid": {"color": "#171796"}},
 "foreground": {"solid": {"color": "#171796"}}
 }],
"categoryAxis": [{
 "show": true,
 "axisScale": "Linear",
 "position": "left",
 "labelColor": {"solid": {"color": "#949494"}},
 "fontFamily": "Arial",
 "fontSize": 8,
 "showAxisTitle": true,
 "titleText": "Add an Axis Title",
 "titleFontSize": 8,
 "titleFontFamily": "Arial",
 "titleColor": {"solid": {"color": "#949494"}},
 "gridLineShow": true,
 "gridLineColor": {"solid": {"color": "#949494"}},
```

```
        "gridLineThickness": 1,
        "gridLineStyle": "Solid"
        }],
    "valueAxis": [{
    "show": true,
    "position": "Left",
    "axisScale": "Linear",
    "labelColor": {"solid": {"color": "#949494"}},
    "fontFamily": "Arial",
    "fontSize": 8,
    "showAxisTitle": true,
    "titleText": "Add an Axis Title",
    "titleFontSize": 8,
    "titleFontFamily": "Arial",
    "titleColor": {"solid": {"color": "#949494"}},
    "gridLineShow": true,
    "gridLineColor": {"solid": {"color": "#949494"}},
    "gridLineThickness": 1,
    "gridLineStyle": "Solid",
    "axisStyle": "ShowTitleOnly"
    }],
    "labels": [{
    "show": true,
    "labelPosition": "InsideEnd",
    "color": {"solid": {"color": "#ffffff"}},
    "fontFamily": "Arial",
    "fontSize": 7,
    "enableBackground": false,
    "backgroundColor": {"solid": {"color": "#d9d9d9"}},
    "backgroundTransparency": 0
    }],
    "plotArea": [{
    "transparency": 0
    }],
    "legend": [{
    "show": true,
    "position": "Bottom",
    "showTitle": true,
    "titleText": "Add a legend title",
    "legendColor": {"solid": {"color": "#949494"}},
    "fontFamily": "Arial",
    "fontSize": 8
    }],
    "lineStyles": [{
    "ShadedArea": true,
    "strokeWidth": 1,
    "lineStyle": "Solid",
    "showMarker": true,
    "markerShape": "Triangle",
    "markerSize": 2,
    "markerColor": {"solid": {"color": "#949494"}}
    }] ,
```

```json
        "slices": [{
        "innerRadiusRatio": 20
        }] ,
        "grid": [{
        "gridVertical": true,
        "gridVerticalColor": {"solid": {"color": "#171796"}},
        "gridVerticalWeight": 1,
        "gridHorizontal": true,
        "gridHorizontalColor": {"solid": {"color": "#171796"}},
        "gridHorizontalWeight": 1,
        "textSize": 10,
        "rowPadding": 2,
        "outlineColor": {"solid": {"color": "#ffffff"}},
        "outlineWeight": 1
        }],
        "columnHeaders": [{
        "fontColor": {"solid": {"color": "#000000"}},
        "backColor": {"solid": {"color": "#E34B20"}},
        "outline": "Frame",
        "alignment": "Left",
        "wordWrap": true
        }],
        "values": [{
        "fontColorPrimary": {"solid": {"color": "#3D4E57"}},
        "backColorPrimary": {"solid": {"color": "#F5F4F0"}},
        "fontColorSecondary": {"solid": {"color": "#3D4E57"}},
        "backColorSecondary": {"solid": {"color": "#E3DFD4"}},
        "outline": "Frame",
        "outline": "Frame",
    "alignment": "Left",
    "wordWrap": true
    }],
    "total": [{
    "totals": true,
    "fontColor": {"solid": {"color": "#000000"}},
    "backColor": {"solid": {"color": "#D1CAB8"}},
    "outline": "Frame"
    }]
    }
   }
  }
  }
}
```

Now that you have built a JSON theme file that can instantly format much (if not most) of a Power BI dashboard, it is worth reminding yourself of what you have created.

The file is nothing but JSON; however, it can be subdivided into three main areas:

- Colors

- Text (textClasses)

- Visuals (visualStyles)

61

This file contains elements that are *cumulative* – in the sense that it applies many different areas of formatting at once.

There is a *hierarchy of styles*. That is, you can apply an initial "top-level" style (for instance, by defining the title element of the textClasses) – and then you can also specify that the titles of charts are formatted differently. These subtleties are explained in the final chapter.

As I mentioned earlier, when defining formatting, you are *not obliged to define every aspect of a formatted element*. In other words, not all "sections" are compulsory, and you can choose which parts of a visual (or a set of visuals that share this common element) you want to format using the theme file.

The order of the elements in the theme file is not important. So you could place the visualStyles element above the textClasses element. What really matters, however, is

- Ensuring that the JSON is syntactically correct.

- Checking that the keywords that define a stylistic element are those expected by Power BI.

- Any text elements (such as font names or choices from popups in Power BI Desktop) contain the values that Power BI expects.

What is more, you can, of course, take the complete "generic" theme file (GenericBaseFormatting.json) and adjust it to use the colors and other settings that you prefer. You can also extend it to add further high-level formatting attributes. You will learn the complete extent of detailed JSON formatting in the next eight chapters.

Hints on Writing JSON Theme Files

JSON theme files can be very dense, can appear cryptic, and can, when you load them, produce the dreaded "Error importing theme file" dialog that is shown in Figure 4-3.

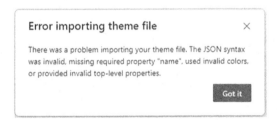

Figure 4-3. Theme file import error

So what can you do to minimize errors when creating theme files?

One of your first actions should be to use your preferred search engine and look for "JSON syntax checker." There are many online resources that allow you to test the structural validity of your JSON. This is particularly useful for checking that the JSON is "well formed" – which means that there are no superfluous (or missing) commas, braces or square brackets, and so on. Some of the tools that allow you to write JSON also incorporate syntax checkers or can add them.

Another key step is to write JSON using a tool that either formats the JSON intelligibly or at the very least allows you to appreciate the structure. By this I mean that being able to appreciate, visually, the levels of nesting and see how pairs of braces match up is key to making your life easier when crafting theme files.

Nothing except loading (or attempting to load) a theme file into Power BI Desktop will test if the actual JSON structure and principal keywords have been typed in correctly, unfortunately. At this level, trial and error seem to be the only way to test that a theme file does what you want it to do.

Sometimes, when a theme file cannot be loaded, you will see an alert like the one shown in Figure 4-4 above the dashboard.

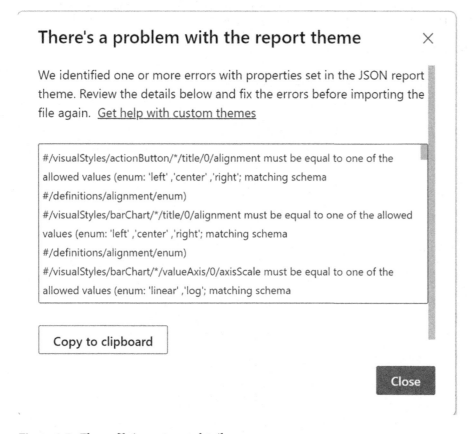

⊗ Unable to import theme. There's a problem with your report theme JSON file. Select More details to troubleshoot the issue. | More details | ✕

Figure 4-4. *Theme file import details error*

In cases like this, the trick is to click the More details button. This will show a dialog something like the one displayed in Figure 4-5.

There's a problem with the report theme   ✕

We identified one or more errors with properties set in the JSON report theme. Review the details below and fix the errors before importing the file again. Get help with custom themes

#/visualStyles/actionButton/*/title/0/alignment must be equal to one of the allowed values (enum: 'left' ,'center' ,'right'; matching schema #/definitions/alignment/enum)
#/visualStyles/barChart/*/title/0/alignment must be equal to one of the allowed values (enum: 'left' ,'center' ,'right'; matching schema #/definitions/alignment/enum)
#/visualStyles/barChart/*/valueAxis/0/axisScale must be equal to one of the allowed values (enum: 'linear' ,'log'; matching schema

| Copy to clipboard |

| Close |

Figure 4-5. *Theme file import error details*

This dialog lists all the errors that prevent the theme file from loading. Most often, this is because an expected element is misspelled. You can now use the information from the dialog to trace the error in your theme file.

■ **Note**   You can also copy the error details and then paste them into a text editor rather than try and read the information in the error dialog.

Conclusion

This chapter took you further into the detail of what constitutes a Power BI theme file. You saw how the JSON in a theme can be built up step by step to define as many – or as few – formatting elements that you want.

While the true potential of JSON themes for Power BI is to define the exact specifications of how each and every visual type will be formatted – down to the most minute level of specific detail – it is not always necessary to push the envelope this far.

Consequently, in this chapter, you saw that JSON themes can be created at a higher, more generic level in order to format a range of dashboard elements without having to describe each element in detail.

Now that you have seen the basics of theme files in action, it is time to move on to an in-depth examination of the precise formatting of every single aspect of Power BI dashboards using JSON. As you might expect, this is a large subject that involves a considerable amount of detail. Consequently, I have broken down the subject matter into eight separate chapters to facilitate understanding. We will start with formatting Power BI dashboard objects.

CHAPTER 5

■ ■ ■

Object Visual Styles

From this point onward in this book, you will be delving into the deepest recesses of JSON theme files. This has the potential to be overwhelming, so I intend to proceed progressively to show you how JSON formatting is largely a process of building up the required knowledge step by step. Moreover, as Power BI themes contain many repetitive pieces of code, you will learn to discern the core elements that are used over and over again in many (if not most) visuals from the code that is specific to each visual. The art is to learn to appreciate not only how JSON is used to define visual formatting but also to build on acquired knowledge and principles to use as a springboard to discover new areas where themes can be applied.

In this chapter, you will see how to write the JSON theme code to define – completely, in the most minute details – the following visuals:

- Textbox

- Image

- Shape

These visuals have the advantage of being less complex than some of the other available visuals, which makes them ideal as the subject of an introduction to more complex and detailed JSON formatting. Moreover, they are not bound to data, which makes them simpler to format – and so to predefine using theme files. An added positive point is that many of their formatting attributes are common to many of the visuals that are available in Power BI. This means that you can learn some of the essential formatting keywords and approaches through the medium of these visuals and then apply the formatting techniques that you have discovered to a whole range of other visuals as you progress through this book.

Each of the three visual types discussed in this chapter has a separate JSON file in the samples folder that contains the code that you can use to define your own theme files. So, let's move on with the journey!

Textbox

To begin your voyage into detailed JSON formatting, the textbox visual is a good place to start. The reason for this is that this particular visual contains, essentially, the set of elements that are common to nearly all visuals. These elements correspond to the formatting that is available in the *General* tab of the format pane. However, the keywords used do not correspond entirely to the sections that appear in the format pane – as you can see in Table 5-1.

You can find the JSON code that formats a textbox in the sample file Attributes_Textbox.json. It is a complete, stand-alone theme file that can be used as a Power BI theme. However, it will only format the color palette, the dashboard background and foreground colors, and, of course, any textboxes that you add to a dashboard.

© Adam Aspin 2023
A. Aspin, *Pro Power BI Theme Creation*, https://doi.org/10.1007/978-1-4842-9633-2_5

The following code snippet provides the JSON code used to pre-format a textbox:

```
{
"name": "Textbox  Attributes",
"dataColors": ["#171796", "#E34B20", "#D1CAB8", "#465437", "#E2EB00", "#615E55", "#00AC98",
"#73268c"],
"background": "#FFFFFF",
"foreground": "#949494",
"visualStyles":
{
"textbox":
{"*" :
 {
 "myTextboxNotes" : "Currently you can add notes like this for visuals",
"title":      [{
    "show": true,
    "text": "Card Title",
    "heading": "Heading2",
    "fontColor": {"solid": {"color": "#ffffff"}},
    "background": {"solid": {"color": "#12239E"}},
    "alignment": "center",
    "fontSize": 10,
    "fontFamily": "Arial",
    "bold": true,
    "italic": false,
    "underline": false
}],
"subTitle":      [{
    "show": true,
    "text": "Textbox Subitle",
    "heading": "Heading3",
    "fontColor": {"solid": {"color": "#ffffff"}},
    "background": {"solid": {"color": "#12239E"}},
    "alignment": "left",
    "fontSize": 9,
    "fontFamily": "Arial",
    "bold": true,
    "italic": false,
    "underline": false,
    "titleWrap": true
        }],
"divider":      [{
    "show": true,
    "color": {"solid": {"color": "#615E55"}},
    "style": "solid",
    "width": 2,
    "ignorePadding": true
        }],
"spacing": [{
    "customizeSpacing": true,
    "spaceBelowTitle": 3,
    "spaceBelowSubTitle": 3,
```

```
        "spaceBelowTitleArea": 2
        }],
"background": [{
    "show": true,
    "color": {"solid": {"color": "#F5F4F0"}},
    "transparency": 0
}],
"lockAspect": [{
"show": false
 }],
"border": [{
    "show": true,
    "color": {"solid": {"color": "#F5F4F0"}},
    "radius": 5
    }],
"dropShadow": [{
    "show": false,
    "color": {"solid": {"color": "#E8E8E8"}},
    "position":"Outer",
    "preset":"TopLeft"
}],
"visualHeaderTooltip":[{
    "type": "Default",
    "titleFontColor":{"solid":{"color":"#E8E8E8"}},
    "text": "Add a tooltip text",
    "fontSize": 11,
    "fontFamily": "Arial Black",
    "background":{"solid":{"color":"#FFFFFF"}},
    "transparency": 30,
    "bold": true,
    "italic": false,
    "underline": false
    }],
"visualHeader": [{
    "show": true,
    "background": {"solid": {"color": "#ffffff"}},
    "border": {"solid": {"color": "#F5F4F0"}},
    "transparency": 10,
    "foreground": {"solid": {"color": "#F5F4F0"}},
    "showVisualInformationButton": false,
    "showVisualWarningButton": false,
    "showVisualErrorButton": false,
    "showDrillRoleSelector": true,
    "showDrillToggleButton": false,
    "showDrillUpButton": false,
    "showDrillDownLevelButton": false,
    "showDrillDownExpandButton": false,
    "showPinButton": false,
    "showFocusModeButton": false,
    "showFilterRestatementButton": false,
    "showSeeDataLayoutToggleButton": false,
    "showOptionsMenu": false,
```

```
    "showTooltipButton": true,
"general": [{
    "responsive": true,
    "altText": "Enter an alternative text"
    }]
  }
 }
}
}
```

In practice, of course, a complete JSON theme file could contain the specification of dozens of visuals. However, I wanted to show you how the code for a visual fits into the overall structure of a theme file, as well as showing you the precise code that you can adjust to define the presentation of a specific visual. You can, of course, take the code for the textbox object and add it to any existing theme file – or replace the definition of the textbox, in whole or in part, by this code in an existing theme file.

So, what are you looking at here?

The first thing to note is that all visuals are defined as part of the *visualStyles* object. That is, the keyword visualStyles is followed by a colon and a pair of curly braces that contain further objects.

The second point is that all visual definitions are contained inside a second object – currently defined as an asterisk. This is nothing more than an empty object reserved for future use. So the actual visual definition is nested inside two outer objects.

The third point is that the definition of a visual is also an object – as you can see, it is contained inside curly braces. In the case of the textbox visual, the keyword that introduces the object is *textbox*. This is highlighted in italics in the preceding JSON code. The nested keywords inside the textbox object that introduce the actual formatting for an element in a textbox are highlighted in italics to draw your attention to how the structure is defined.

The final important point is that the JSON for a specific visual consists of mapping the Power BI Desktop format pane section elements to the JSON keyword that you have to use to specify the section element to format in the theme file.

Table 5-1 describes the JSON keyword that corresponds to each section in the Power BI Desktop format pane.

Table 5-1. *Textbox Formatting Elements*

Keyword	Formatting Section
padding	Properties/Padding
Title	Title/Title
subTitle	Title/Subtitle
divider	Title/Divider
spacing	Title/Spacing
background	Effects/Background
border	Effects/Visual border
dropShadow	Effects/Shadow
visualHeader	Header Icons
lockAspect	Lock aspect
visualHeaderTooltip	Header Icons/Help tooltip

You can see that a textbox actually contains no element that is specific to a textbox. All the key/value pairs actually describe elements that are found in most visuals. The reason the JSON for a textbox describes nothing other than these seven elements is because the text inside a textbox is essentially formatted specifically for each textbox that you create.

To begin the detailed explanation of items that can be formatted, the starting point is the on/off switch in the format pane. This is common to many elements and is always represented by the following JSON:

```
"show": true
```

or (to switch it off)

```
"show": false
```

Other elements are, necessarily, more specific. Let's look at these in turn.

The *title* keyword that describes the title section uses the keywords shown in Table 5-2.

Table 5-2. Keyword Mapping for Titles

Element	Keyword
Title	Show
Title text	Text
Heading	Heading
Font	fontFamily
(Font size)	fontSize
(Bold)	Bold
(Italic)	Italic
(Underline)	Underline
Text color	fontColor
Background color	background
Alignment	Alignment
Text wrap	textWrap

The *subtitle* keyword that describes the background section uses the keywords shown in Table 5-3.

Table 5-3. *Keyword Mapping for Subtitle*

Element	Keyword
Title	Show
Title text	Text
Heading	Heading
Font	fontFamily
(Font size)	fontSize
(Bold)	Bold
(Italic)	Italic
(Underline)	Underline
Text color	fontColor
Background color	Background
Alignment	Alignment
Text wrap	titleWrap

The *divider* keyword that describes the background section uses the keywords shown in Table 5-4.

Table 5-4. *Keyword Mapping for Divider*

Element	Keyword
Show	Show
Color	Color
Style	Style
Width	Width
Ignore padding	ignorePadding

The *spacing* keyword that describes the background section uses the keywords shown in Table 5-5.

Table 5-5. *Keyword Mapping for Spacing*

Element	Keyword
Customize spacing	customizeSpacing
Space below title	spaceBelowTitle
Space below subtitle	spaceBelowSubtitle
Space below title area	spaceBelowTitleArea

The *background* keyword that describes the background section uses the keywords shown in Table 5-6.

***Table 5-6.** Keyword Mapping for Background*

Element	Keyword
Background	Show
Color	Color
Transparency	Transparency

The *lockAspect* keyword that describes the lock aspect section uses the keyword shown in Table 5-7.

***Table 5-7.** Keyword Mapping for Lock Aspect*

Element	Keyword
Lock aspect	Show

The *border* keyword that describes the border section uses the keywords shown in Table 5-8.

***Table 5-8.** Keyword Mapping for Border*

Element	Keyword
Border	show
Transparency	transparency
Rounded corners	Radius

The *shadow* keyword that describes the shadow section uses the keywords shown in Table 5-9.

***Table 5-9.** Keyword Mapping for Shadow*

Element	Keyword
Shadow	show
Color	color
Offset	position
Position	preset

The *visualHeader* keyword that describes the visual header section uses the keywords shown in Table 5-10.

Table 5-10. *Keyword Mapping for the Visual Header*

Element	Keyword
Background	Background
Border	Border
Transparency	Transparency
Icon	Foreground
Visual information icon	showVisualInformationButton
Visual warning icon	showVisualWarningButton
Visual error icon	showVisualErrorButton
Drill on dropdown	showDrillRoleSelector
Drill down icon	showDrillToggleButton
Drill up icon	showDrillUpButton
Show next level icon	showDrillDownLevelButton
Expand to next level icon	showDrillDownExpandButton
Pin icon	showPinButton
Focus mode icon	showFocusModeButton
Filter icon	showFilterRestatementButton
See data layout icon	showSeeDataLayoutToggleButton
More options icon	showOptionsMenu
Visual header tooltip icon	showTooltipButton
Smart Narrative	smartNarrativeButton

The *general* keyword that describes the Properties/Advanced options and Alt Text sections uses the keywords shown in Table 5-11.

Table 5-11. *Keyword Mapping for the Properties/Advanced Options and Alt Text Sections*

Element	Keyword
Responsive	responsive
Alt Text	altText

The *padding* keyword that describes the Properties/Padding section uses the keywords shown in Table 5-12.

Table 5-12. *Keyword Mapping for the Padding Section*

Element	Keyword
Left	Left
Right	Right
Top	Top
Bottom	Bottom

Something that you really need to understand is that a theme file will require, for some attributes, that you specify an element that corresponds to a selection in a popup list in Power BI Desktop. An example is the alignment of titles that is used in visual titles as well as column headers and certain other attributes. If you look at the preceding JSON, you can see the following line of code for the title and subtitle:

```
"alignment": "Center"
```

As you have probably guessed, this corresponds to selecting the icon that indicates *center* in the Power BI Desktop format pane.

When setting key/value pairs like this, all you need to know is *exactly* what value you can use in the JSON. In some cases, these values are a direct match to the values in the popup that is used in Power BI Desktop. Unfortunately, this is not always the case. So I will always provide the keyword mapping in the course of the remaining chapters.

The *alignment* keyword that is used in the title and subtitle sections needs one of the attributes shown in Table 5-13.

Table 5-13. *Title Alignment Options*

Value	Formatting Popup Option
Auto	Auto
Left	Left
Center	Center
Right	Right

The *position* keyword that is used in the drop shadow section needs one of the attributes shown in Table 5-14.

Table 5-14. *Shadow Position Options*

Value	Formatting Popup Option
Outer	Outside
Inner	Inside

The *preset* keyword that is used in the drop shadow section needs one of the attributes shown in Table 5-15.

Table 5-15. *Shadow Preset Options*

Value	Formatting Popup Option
BottomRight	Bottom Right
Bottom	Bottom
BottomLeft	Bottom Left
Center	Center
Left	Left
TopRight	Top Right
Top	Top
TopLeft	Top Left
Custom	Custom

As so many of the detailed values that are defined in the JSON are used identically across multiple visuals and the same way inside the same visual, I will not explain each one exhaustively over and over again when each visual is discussed.

■ **Note** It is *really important* when entering a value in the theme file that corresponds to a specific formatting option that you use *exactly the spelling and capitalization* that is expected by Power BI – and given in the reference table.

Image

The next visual that I suggest looking at is the image. This is because the image visual only contains two elements over and above the standard elements that you saw for the textbox visual. Indeed, these options are only found in a couple of other elements in Power BI – and they are the action and the *scaling* that you can apply. These are defined using the *visualLink* and *imageScaling* keywords.

Image formatting (as you can see in the following JSON snippet) is introduced by the *image* keyword.

Table 5-16 describes the JSON keyword that corresponds to each formatting section in Power BI Desktop as far as images are concerned.

Table 5-16. *Image Formatting Elements*

Keyword	Formatting Section
general	General
imageScaling	Scaling
Action	visualLink
padding	Properties/Padding
title	Title/Title
subTitle	Title/Subtitle
divider	Title/Divider
spacing	Title/Spacing
background	Effects/Background
border	Effects/Visual border
dropShadow	Effects/Shadow
visualHeader	Header Icons
lockAspect	Lock aspect
visualHeaderTooltip	Header Icons/Help tooltip
title	Title
background	Background
border	Border
dropShadow	Shadow
visualHeader	Visual header

Here, then, is the JSON code that you can use to format an image. You can find this file in the sample files – it is Attributes_Image.json. Once again, the new – or essential – keywords are highlighted in italics. Please note that from now on I will only be showing the JSON that is specific to the visual under discussion. The common code that encapsulates the specific JSON code will not be displayed – even if it is in the corresponding sample file.

```
"image":
  {"*":
    {
    "general": [{... code not shown ...}]
    "imageScaling": [{
                      "imageScalingType": "Fit"
                  }],
    "visualLink": [{
              "show": true,
              "type": "Bookmark",
```

```
                "tooltip": "Tooltip text here"
             }],
    "title": [{... code not shown ...}],
    "subTitle": [{... code not shown ...}],
    "divider": [{... code not shown ...}],
    "spacing": [{... code not shown ...}],
    "background": [{... code not shown ...}],
    "lockAspect": [{... code not shown ...}],
    "border": [{... code not shown ...}],
    "dropShadow": [{... code not shown ...}],
    "visualHeader": [{... code not shown ...}]
    "visualHeaderTooltip": [{... code not shown ...}]
    "padding": [{... code not shown ...}]
    }
}
```

The *imageScalingType* keyword in the JSON that is used in the imageScaling section needs one of the attributes shown in Table 5-17.

Table 5-17. *Label Position Options*

Value	Formatting Popup Option
Normal	Normal
Fit	Fit
Fill	Fill

The *visualLink* keyword in the JSON that is used in the imageScaling section needs one of the attributes shown in Table 5-18.

Table 5-18. *Visual Link Options*

Value	Formatting Popup Option
Back	Back
Bookmark	Bookmark
Page navigation	PageNavigation
Q&A	Qna
Web Url	WebUrl
Apply all slicers	applyAllSlicers
Clear all slicers	Clear all slicers

Shape

To move on (and to conclude the chapter), the next visual to look at is the shape visual. The keyword that specifies shape formatting is *shape*. The format of this visual is defined using the following keywords:

- shape
- fill
- rotation
- outline
- text
- shadow
- glow

It follows that you need to see how these formatting elements can be configured using JSON in a theme file.

Table 5-19 describes the JSON keyword that corresponds to each formatting section in Power BI Desktop.

Table 5-19. *Shape Formatting Elements*

Keyword	Formatting Section
general	General
shape	Shape
rotation	Rotation
fill	Style/Fill
outline	Style/Border
text	Style/Text
shadow	Style/Shadow
glow	Style/Glow
padding	Properties/Padding
title	Title/Title
subTitle	Title/Subtitle
divider	Title/Divider
spacing	Title/Spacing
background	Effects/Background
border	Effects/Visual border
dropShadow	Effects/Shadow
visualHeader	Header Icons/Icons Header Icons/Colors
lockAspect	Lock aspect
visualHeaderTooltip	Header Icons/Help tooltip

So, for the shape element, the code would look like the following piece of JSON (this code can be found in the sample file Attributes_Shape.json):

```
"basicShape":
  {"*" :
    {
    "general": [{... code not shown ...}]
    "shape":      [{
        "tileShape": "line",
        "roundedge": 0

        }],
    "rotation": [{
        "angle": 0,
        "shapeAngle": 0,
        "textAngle": 0
        }],
    "fill":      [{
        "show": true,
        "fillColor": {"solid": {"color": "#FFFFFF"}},
        "transparency": 0
          }],
    "rotation":  [{
        "angle": 0
        }],
    "outline":  [{
        "show": false,
        "lineColor": {"solid": {"color": "#F5F4F0"}},
        "transparency": 0,
        "weight": 1,
        "roundedge": 10
        }],
    "text":  [{
        "show": true,
        "text": "Shape Text",
        "fontFamily": "Arial",
        "preserveWhitespace": false,
        "bold": true,
        "italic": false,
        "underline": false,
        "fontColor": {"solid": {"color": "#ffffff"}},
        "horizontalAlignment": "center",
        "verticalAlignment": "middle",
        "topMargin": 2,
        "bottomMargin": 2,
        "leftMargin": 2,
        "rightMargin": 2
        }],
    "shadow":  [{
        "show": true,
        "color": {"solid": {"color": "#FFFFFF"}},
```

```
            "transparency": 0,
            "shadowBlur": 0,
            "shadowPositionPreset": "bottomRight"
            }],
      "glow":   [{
            "show": true,
            "color": {"solid": {"color": "#FFFFFF"}},
            "transparency": 0,
            "blur": 0
            }],
      "visualLink": [{... code not shown ...}],
      "title": [{... code not shown ...}],
      "subTitle": [{... code not shown ...}],
      "divider": [{... code not shown ...}],
      "spacing": [{... code not shown ...}],
      "background": [{... code not shown ...}],
      "lockAspect": [{... code not shown ...}],
      "border": [{... code not shown ...}],
      "dropShadow": [{... code not shown ...}],
      "visualHeader": [{... code not shown ...}]
      "visualHeaderTooltip": [{... code not shown ...}]
      "padding": [{... code not shown ...}]
      }
}
```

The *shape* keyword that describes the Shape section uses the keywords shown in Table 5-20.

Table 5-20. *Keyword Mapping for Shape Elements*

Element	Keyword
Shape	tileShape
Rounded Corners	roundEdge

The *fill* keyword that describes the Style/Fill section uses the keywords shown in Table 5-21.

Table 5-21. *Keyword Mapping for Fill Elements*

Element	Keyword
Fill	Show
Fill color	fillColor
Transparency	transparency

The *rotation* keyword that describes the rotation section uses the keywords shown in Table 5-22.

Table 5-22. *Keyword Mapping for Rotation*

Element	Keyword
Rotation/All	Rotation
Rotation/Shape	shapeAngle
Rotation/Text	textAngle

The *outline* keyword that describes the Style/Border section uses the keywords shown in Table 5-23.

Table 5-23. *Keyword Mapping for Border*

Element	Keyword
Border	Show
Color	fillColor
Width	Weight
Transparency	Transparency

The *text* keyword that describes the Style/Text section uses the keywords shown in Table 5-24.

Table 5-24. *Keyword Mapping for Text*

Element	Keyword
Text	Show
Text	Text
Font	fontFamily
(Font size)	fontSize
(Bold)	Bold
(Italic)	Italic
(Underline)	Underline
Font color	fontColor
Horizontal alignment	horizontalAlignment
Vertical alignment	verticalAlignment
(Top margin)	topMargin
(Bottom margin)	bottomMargin
(Left margin)	leftMargin
(Right margin)	rightMargin

The *shadow* keyword that describes the Style/Shadow section uses the keywords shown in Table 5-25.

Table 5-25. *Keyword Mapping for Shadow*

Element	Keyword
Shadow	Show
Color	Color
Transparency	Transparency
Blur	shadowBlur
Position	shadowPositionPreset

The *glow* keyword that describes the Style/Glow section uses the keywords shown in Table 5-26.

Table 5-26. *Keyword Mapping for Glow*

Element	Keyword
Shadow	Show
Color	Color
Transparency	Transparency
Blur	shadowBlur

The *tileShape* keyword in the JSON that is used in the shape section needs one of the attributes shown in Table 5-27.

Table 5-27. *tileShape Options*

Value	Formatting Popup Option
Arrow	Arrow
Chevron arrow	arrowChevron
Pentagon arrow	arrowPentagon
Heart	Heart
Hexagon	Hexagon
Line	Line
Octagon	Octagon
Oval	Oval
Parallelogram	Parallelogram
Pentagon	Pentagon
Pill	Pill
Rectangle	Rectangle
Rounded rectangle	roundedRectangle
Snipped tab, top right	tabCutCorner
Snipped tab, both top	tabCutTopCorners
Rounded tab, top right	tabRoundCorner
Rounded tab, both top	tabRoundTopCorners
Speech bubble	speechbubbleRectangle
Trapezoid	Trapezoid
Isosceles triangle	triangleIsoc
Right rectangle	triangleRight

■ **Note** At the time of going to press, the fill, outline, shadow, and glow keywords are not working. However, I have extrapolated the required JSON from the underlying .Pbix file in the hope that this will be working by the time that this book is published.

Conclusion

This chapter introduced you to how you can prepare and standardize the in-depth formatting of visuals using JSON theme files in Power BI. Specifically, you saw three of the key visuals in Power BI (textbox, shape, and image) and how to format them.

Each visual is defined by a JSON keyword that, in turn, contains any of the format attributes (corresponding to the sections of the Power BI Desktop format pane for this visual).

Some attributes are simple and only require a value to be defined; others are nested objects and/or arrays.

However, the art and science of creating and modifying JSON theme files will always be to find which JSON keyword maps to which Power BI formatting attribute and which values you can use to define how this attribute is to be defined as standard.

Now that you have seen a number of core elements, it is time to move on to looking at different groups of visuals, beginning with card and table visual types. These are covered in the next chapter.

CHAPTER 6

■ ■ ■

Card and Table Visual Styles

To extend your knowledge of JSON formatting, it is time to move on to a few of the core data-driven visuals. A useful starter set to look at is

- Card

- Multi-row card

- Table

- Matrix

I am handling this group of visuals in a specific order, as I prefer to start with the simplest and move on to progressively more complex visuals in the course of the chapter. This should also help you to understand how to build up your knowledge as JSON themes are defined using repeated – or largely similar – elements.

Each of the four visuals discussed in this chapter has a separate JSON file in the samples folder that contains the code that you can use to define your own theme files.

Card

One of the simplest visuals (which means mercifully simple JSON in the theme file) is the card visual. The keyword that introduces card formatting attributes is *card*.

Table 6-1 describes the JSON keyword that corresponds to each formatting section in Power BI Desktop.

© Adam Aspin 2023
A. Aspin, *Pro Power BI Theme Creation*, https://doi.org/10.1007/978-1-4842-9633-2_6

Table 6-1. *Card Formatting Elements*

Keyword	Formatting Section
General	General
labels	Data label
categoryLabels	Category label
wordwrap	Wordwrap
title	Title
subTitle	Subtitle
divider	Divider
spacing	Spacing
background	Background
lockAspect	Lock aspect
border	Border
dropShadow	Shadow
visualHeader	Header icons
visualHeaderTooltip	Header icons
visualTooltip	Tooltips

The following code snippet gives you the core elements used to specify standard card formatting. The sample file Attributes_Card.json contains all the elements that can be set for a card in a theme file.

```
"card":
  {"*":
    {
    "general": [{... code not shown ...}]
    "labels":   [{
         "color": {"solid": {"color": "#12239E"}},
         "labelDisplayUnits": 1000,
         "labelPrecision": 2,
         "fontSize": 36,
         "fontFamily": "Arial",
         "preserveWhitespace": false,
        "bold": true,
        "italic": false,
        "underline": false
        }],
    "categoryLabels": [{
         "show": true,
         "color": {"solid": {"color": "#12239E"}},
         "fontSize": 24,
         "fontFamily": "Arial",
         "bold": true,
```

```
            "italic": false,
            "underline": false
            }],
    "wordwrap": [{
            "show": true
            }],
    "title": [{... code not shown ...}],
    "subTitle": [{... code not shown ...}],
    "divider": [{... code not shown ...}],
    "spacing": [{... code not shown ...}],
    "background": [{... code not shown ...}],
    "lockAspect": [{... code not shown ...}],
    "border": [{... code not shown ...}],
    "dropShadow": [{... code not shown ...}],
    "visualHeader": [{... code not shown ...}]
    "visualHeaderTooltip": [{... code not shown ...}]
    "visualTooltip": [{... code not shown ...}]
    "padding": [{... code not shown ...}]
    }
}
```

The *labels* keyword that describes the data label section uses the keywords shown in Table 6-2.

Table 6-2. *Keyword Mapping for Data Labels*

Element	Keyword
Color	Color
Display units	labelDisplayUnits
Value decimal places	labelPrecision
Text size	fontSize
Font family	fontFamily
Source spacing	preserveWhitespace
(Bold)	Bold
(Italic)	Italic
(Underline)	Underline

The *categoryLabels* keyword that describes the category labels section uses the keywords shown in Table 6-3.

Table 6-3. *Keyword Mapping for Category Labels*

Element	Keyword
Category labels	Show
Color	Color
Text size	fontSize
Font family	fontFamily
(Bold)	Bold
(Italic)	Italic
(Underline)	Underline

The *wordWrap* keyword that describes the word wrap section uses the keyword shown in Table 6-4.

Table 6-4. *Keyword Mapping for Word Wrap*

Element	Keyword
Word wrap	Show

The *visualTooltip* keyword that describes the tooltip section uses the keywords shown in Table 6-5.

Table 6-5. *Keyword Mapping for Tooltips*

Element	Keyword
Tooltip	show
Label color	titleFontColor
Value color	valueFontColor
Text size	fontSize
Font family	fontFamily
(Bold)	Bold
(Italic)	italic
(Underline)	Underline
Background color	background
Transparency	transparency

Multi-row Card

Multi-row cards are in many ways an extension of the card theme, but are nonetheless sufficiently different to require a set of separate formatting elements.

The keyword that specifies multi-row card formatting attributes is *multiRowCard*.

Table 6-6 outlines the JSON keywords that you can use in multi-row card formatting.

Table 6-6. *Multi-row Card Formatting Elements*

Keyword	Formatting Section
General	General
dataLabels	Data labels
categoryLabels	Category labels
cardTitle	Card title
Card	Card
Title	Title
subTitle	Subtitle
divider	Divider
spacing	Spacing
Background	Background
lockAspect	Lock aspect
Border	Border
dropShadow	Shadow
visualHeader	Header icons
visualHeaderTooltip	Header icons
visualTooltip	Tooltips

The complete JSON required to format a multi-row card can be found in the sample file Attributes_MultiRowCard.json. You can see the elements that are specific to this visual type in the following code snippet:

```
"multiRowCard":
  {"*":
    {
    "general": [{... code not shown ...}]
    "dataLabels": [{
        "color": {"solid": {"color": "#12239E"}},
        "fontSize": 10,
        "fontFamily": "Arial",
        "bold": true,
```

```
            "italic": false,
            "underline": false

        }],
    "categoryLabels":   [{
        "show": true,
        "color": {"solid": {"color": "#12239E"}},
        "fontSize": 20,
        "fontFamily": "Arial",
        "bold": true,
        "italic": false,
        "underline": false

        }],
    "cardTitle":    [{
        "fontColor": {"solid": {"color": "#ffffff"}},
        "fontSize": 16,
        "fontFamily": "Arial",
        "bold": true,
        "italic": false,
        "underline": false

        }],
    "card":    [{
        "outline": "BottomOnly",
        "outlineColor": {"solid": {"color": "#12239E"}},
        "outlineWeight": 3,
        "barShow": true,
        "barColor": {"solid": {"color": "#12239E"}},
        "barWeight": 5,
        "cardPadding": 15,
        "cardBackground": {"solid": {"color": "#F5F4F0"}}
        }],
    "title": [{... code not shown ...}],
    "subTitle": [{... code not shown ...}],
    "divider": [{... code not shown ...}],
    "spacing": [{... code not shown ...}],
    "background": [{... code not shown ...}],
    "lockAspect": [{... code not shown ...}],
    "border": [{... code not shown ...}],
    "dropShadow": [{... code not shown ...}],
    "visualHeader": [{... code not shown ...}],
    "visualHeaderTooltip": [{... code not shown ...}]
    "visualTooltip": [{... code not shown ...}]
    "padding": [{... code not shown ...}]
    }
}
```

The *dataLabels* keyword that describes the data labels section uses the keywords shown in Table 6-7.

Table 6-7. *Keyword Mapping for Data Labels*

Element	Keyword
Color	Color
Text size	fontSize
Font family	fontFamily
(Bold)	Bold
(Italic)	Italic
(Underline)	Underline

The category labels section for a multi-row card can be found in Table 6-3.
The *cardTitle* keyword that describes the card title section uses the keywords shown in Table 6-8.

Table 6-8. *Keyword Mapping for Card Titles*

Element	Keyword
Color	Fontcolor
Text size	fontSize
Font family	fontFamily
(Bold)	Bold
(Italic)	Italic
(Underline)	Underline

The *card* keyword that describes the card section uses the keywords shown in Table 6-9.

Table 6-9. *Keyword Mapping for Card Elements*

Element	Keyword
Outline	Outline
Outline color	outlineColor
Outline weight	outlineWeight
Show bar	barShow
Bar color	barColor
Bar thickness	barWeight
Padding	cardPadding
Background	cardBackground

The *outline* keyword in the JSON that is used in the card section needs one of the attributes shown in Table 6-10.

Table 6-10. *Card Outline Style Options*

Value	Formatting Popup Option
None	None
BottomOnly	Bottom only
TopOnly	Top only
LeftOnly	Left only
RightOnly	Right only
TopBottom	Top + bottom
LeftRight	Left + right
Frame	Frame

■ **Note**   Once again, the spelling of the formatting has to be exactly as given in the preceding tables. If you add spaces or do not respect the capitalization, then the table style will not be applied. The default value will be applied instead.

Table

As one of the most widely used visuals in Power BI dashboards, you will doubtless need to know the details (and intricacies) of how to format tables using JSON theme files.

To begin the overview, the essential keyword for defining table formatting attributes is *tableEx*.

The table elements that you can format using a theme file (over and above the standard visual elements) are

- Style
- Grid
- Column headers
- Values
- Total
- Tooltip

These elements are shown in italics in the following code snippet.

Table 6-11 outlines the JSON keywords that you can use in table visual formatting and the section that corresponds to each keyword in the Power BI Desktop format pane.

Table 6-11. *Table Keywords*

Keyword	Formatting Popup Option
General	General
stylePreset	Style
Grid	Grid
columnHeaders	Column headers
Values	Values
Total	Total
Title	Title
subtitle	Subtitle
Divider	Divider
Spacing	Spacing
lockAspect	Lock aspect
Border	Border
dropShadow	Shadow
visualHeader	Header icons
visualHeaderTooltip	Header icons
visualTooltip	Tooltips

A complete theme file (which also contains the formatting for standard aspects such as background, title, lock aspect, border, shadow, and the visual header) is available in the sample data files as Attributes_Table.json.

As you can see in the following, this is a somewhat longer piece of JSON. After all, it has to define a larger set of stylistic options. However, it follows the standards that you have seen so far in smaller JSON snippets and consequently should be comprehensible. The trick, I find, is to consider it section by section, rather than trying to take it all in at once.

The following code snippet provides the JSON code used to pre-format a table:

```
"tableEx":
  {"*":
    {
    "general": [{... code not shown ...}]
    "stylePreset": [{"name": "ContrastAlternatingRows"}],
    "grid":     [{
        "gridVertical": false,
        "gridVerticalColor": {"solid": {"color": "#F5F4F0"}},
        "gridVerticalWeight": 1,
        "gridHorizontal": false,
        "gridHorizontalColor": {"solid": {"color": "#F5F4F0"}},
        "gridHorizontalWeight": 1,
        "rowPadding": 1,
```

```
      "outlineColor": {"solid": {"color": "#ffffff"}},
      "outlineWeight": 1,
      "outlineStyle": 1,
      "textSize": 10,
      "imageHeight": 60
      }],
   "columnHeaders": [{
      "fontColor": {"solid": {"color": "#ffffff"}},
      "backColor": {"solid": {"color": "#118Dff"}},
      "outline": "TopBottom",
      "autoSizeColumnWidth": true,
      "fontFace": "Arial",
      "fontSize": 8,
      "alignment": "Left",
      "wordWrap": true,
      "bold": true,
      "italic": false,
      "underline": false
      }],
   "values": [{
      "fontColorPrimary": {"solid": {"color": "#949494"}},
      "backColorPrimary": {"solid": {"color": "#E8E8E8"}},
      "fontColorSecondary": {"solid": {"color": "#808080"}},
      "backColorSecondary": {"solid": {"color": "#D5D5D5"}},
      "outline": "Frame",
      "urlIcon": true,
      "wordWrap": true,
      "fontFamily": "Arial",
      "fontSize": 8,
      "bold": true,
      "italic": false,
      "underline": false

      }],
   "total":    [{
      "totals": true,
      "label": "TOTAL",
      "fontColor": {"solid": {"color": "#000000"}},
      "backColor": {"solid": {"color": "#118DFF"}},
      "outline": "Frame",
      "fontFamily": "Arial",
      "fontSize": 10,
      "bold": true,
      "italic": false,
      "underline": false
      }],
```

```
    "title": [{... code not shown ...}],
    "subTitle": [{... code not shown ...}],
    "divider": [{... code not shown ...}],
    "spacing": [{... code not shown ...}],
    "background": [{... code not shown ...}],
    "lockAspect": [{... code not shown ...}],
    "border": [{... code not shown ...}],
    "dropShadow": [{... code not shown ...}],
    "visualHeader": [{... code not shown ...}]
    "visualHeaderTooltip": [{... code not shown ...}]
    "visualTooltip": [{... code not shown ...}]
    "padding": [{... code not shown ...}]
    }
}
```

The *grid* keyword that describes the grid section uses the keywords shown in Table 6-12.

Table 6-12. *Keyword Mapping for the Grid*

Element	Keyword
Vertical grid	gridVertical
Vertical grid color	gridVerticalColor
Vertical grid thickness	gridVerticalWeight
Horizontal grid	gridHorizontal
Horizontal grid color	gridHorizontalColor
Horizontal grid thickness	gridHorizontalWeight
Row padding	rowPadding
Outline color	outlineColor
Outline weight	outlineWeight
Text size	textSize
Image height	imageHeight

The *columnHeaders* keyword that describes the column headers section uses the keywords shown in Table 6-13.

Table 6-13. *Keyword Mapping for Column Headers*

Element	Keyword
Font color	fontColor
Background color	backColor
Outline	outline
Auto-size column width	autoSizeColumnWidth
Font family	fontFace
Text size	fontSize
Alignment	alignment
Word wrap	wordWrap
(Bold)	bold
(Italic)	italic
(Underline)	underline

The *values* keyword that describes the values section uses the keywords shown in Table 6-14.

Table 6-14. *Keyword Mapping for Values*

Element	Keyword
Font color	fontColorPrimary
Background color	backColorPrimary
Alternate font color	fontColorSecondary
Alternate background color	backColorSecondary
Outline	Outline
URL icon	urlIcon
Word wrap	wordWrap
Font family	fontFamily
Text size	fontSize
(Bold)	Bold
(Italic)	Italic
(Underline)	Underline

The *total* keyword that describes the total section uses the keywords shown in Table 6-15.

Table 6-15. *Keyword Mapping for Totals*

Element	Keyword
Totals	Totals
Label	Label
Font color	fontColor
Background color	backColor
Outline	Outline
Font family	fontFamily
Text size	fontSize
(Bold)	Bold
(Italic)	Italic
(Underline)	Underline

The *stylePreset* keyword in the JSON that is used in the overall table presentation style needs one of the attributes shown in Table 6-16.

Table 6-16. *Table Style Options*

Value	Formatting Popup Option
Default	Default
None	None
Minimal	Minimal
BoldHeader	Bold header
AlternatingRows	Alternating rows
ContrastAlternatingRows	Contrast alternating rows
FlashyRows	Flashy rows
BoldHeaderFlashyRows	Bold header flashy rows
Sparse	Sparse
Condensed	Condensed

The *outlineStyle* keyword in the JSON that is used to define border display needs one of the attributes shown in Table 6-17.

Table 6-17. *Outline Style Options*

Value	Formatting Popup Option
1	Top
2	Right
4	Bottom
8	Left

These values can be combined (up to a total of 15) to apply multiple borders. So, for instance, using 6 as the outlineStyle specification will add right and bottom borders.

Tooltip options are described in Table 6-5.

The *outline* keyword that describes both the columnHeaders and values sections uses the keywords shown in Table 6-10 in the previous section.

The following elements are *not* currently configurable using JSON themes:

- Field formatting

- Conditional formatting

Matrix

As you might expect, the JSON that defines the formatting for a matrix is largely similar to – indeed is built on and extends – the JSON for table formatting.

The elements that are the same as those found in the JSON for a table are

- Style

- Grid

Other elements (values and column headers, for instance) are slightly different. If you look at the complete JSON for these elements, you can compare and contrast the two code snippets. The subtotals and totals elements are sufficiently different to be worth looking at in detail.

The keyword that introduces matrix formatting attributes is *pivotTable*.

The keywords that are used when formatting a matrix are given in Table 6-18 along with the section that corresponds to each keyword in the Power BI Desktop format pane.

Table 6-18. *Matrix Keywords*

Keyword	Formatting Popup Option
general	General
stylePreset	Style
grid	Grid
columnHeaders	Column headers
rowHeaders	Row headers
values	Values
subTotals	Subtotals
RowTotal	Row grand total
columnTotal	Column grand total
title	Title
subTitle	Subtitle
divider	Divider
spacing	Spacing
lockAspect	Lock aspect
border	Border
dropShadow	Shadow
visualHeader	Header icons
visualHeaderTooltip	Header icons
visualTooltip	Tooltips

The following code snippet shows how you can format a matrix in a JSON file. A full piece of JSON that contains all the elements that you can preset is in the sample files as Attributes_Matrix.json.

```
"pivotTable":
  {"*" :
    {
    "general": [{... code not shown ...}]
    "stylePreset":  [{"name": "ContrastAlternatingRows"}],
    "grid": [{
        "gridVertical": false,
        "gridVerticalColor": {"solid": {"color": "#E8E8E8"}},
        "gridVerticalWeight": 1,
        "gridHorizontal": false,
        "gridHorizontalColor": {"solid": {"color": "#E8E8E8"}},
        "gridHorizontalWeight": 1,
        "rowPadding": 1,
        "outlineColor": {"solid": {"color": "#ffffff"}},
        "outlineWeight": 1,
```

```
            "textSize": 10,
            "imageHeight": 60
    }],
    "columnHeaders": [{
            "fontColor": {"solid": {"color": "#ffffff"}},
            "backColor": {"solid": {"color": "#12239E"}},
            "outline": "TopBottom",
            "autoSizeColumnWidth": true,
            "fontFace": "Arial",
            "fontSize": 8,
            "alignment": "Left",
            "titleAlignment": "Center",
            "wordWrap": true,
            "bold": true,
            "italic": false,
            "underline": false
    }],
    "rowHeaders": [{
            "fontColor": {"solid": {"color": "#ffffff"}},
            "backColor": {"solid": {"color": "#118DFF"}},
            "outline": "TopBottom",
            "stepped": true,
            "steppedLayoutIndentation": 14,
            "urlIcon": true,
            "wordWrap": true,
            "fontFamily": "Arial",
            "fontSize": 8,
            "titleAlignment": "Center",
            "showExpandCollapseButtons": true,
            "expandCollapseButtonsColor": {"solid": {"color": "#ffffff"}},
            "expandCollapseButtonsSize": 12,
            "bold": true,
            "italic": false,
            "underline": false
    }],
    "values": [{
            "fontColorPrimary": {"solid": {"color": "#3D4E57"}},
            "backColorPrimary": {"solid": {"color": "#F5F4F0"}},
            "fontColorSecondary": {"solid": {"color": "#3D4E57"}},
            "backColorSecondary": {"solid": {"color": "#E3DFD4"}},
            "bandedRowHeaders": true,
            "valuesOnRow": false,
            "outline": "Frame",
            "urlIcon": true,
            "wordWrap": true,
            "fontFamily": "Arial",
            "fontSize": 8,
            "bold": true,
            "italic": false,
            "underline": false
    }],
```

```
  "subTotals": [{
        "$id": "Row",
        "fontColor": {"solid": {"color": "#ffffff"}},
        "fontSize": 10,
        "fontFamily": "Arial",
        "bold": true,
        "italic": false,
        "underline": false,
        "applyToHeaders": false,
        "backColor": {"solid": {"color": "#ffffff"}}
  },
  {

        "$id": "Column",
        "fontColor": {"solid": {"color": "#ffffff"}},
        "fontSize": 10,
        "fontFamily": "Arial",
        "bold": true,
        "italic": false,
        "underline": false,
        "applyToHeaders": false,
        "backColor": {"solid": {"color": "#ffffff"}}
  },
  {

        "rowSubTotals": true,
        "columnSubTotals": true,
        "rowSubTotalsPosition": "Bottom",
        "perRowLevel": false,
        "perColumnLevel": false,
        "rowSubtotalsLabel": "Total",
        "columnSubtotalsLabel": "Total"
  }],
  "rowTotal":      [{
        "fontColor": {"solid": {"color": "#3D4E57"}},
        "fontFamily": "Arial",
        "backColor": {"solid": {"color": "#D1CAB8"}},
        "applyToHeaders": true,
        "fontSize": 10,
        "bold": true,
        "italic": false,
        "underline": false
  }],
  "columnTotal":      [{
        "fontColor": {"solid": {"color": "#3D4E57"}},
        "fontFamily": "Arial",
        "backColor": {"solid": {"color": "#D1CAB8"}},
        "applyToHeaders": true,
        "fontSize": 10,
        "bold": true,
        "italic": false,
        "underline": false
        }],
```

```
    "title": [{... code not shown ...}],
    "subTitle": [{... code not shown ...}],
    "divider": [{... code not shown ...}],
    "spacing": [{... code not shown ...}],
    "background": [{... code not shown ...}],
    "lockAspect": [{... code not shown ...}],
    "border": [{... code not shown ...}],
    "dropShadow": [{... code not shown ...}],
    "visualHeader": [{... code not shown ...}]
    "visualTooltip": [{... code not shown ...}]
    "visualHeaderTooltip": [{... code not shown ...}]
    "padding": [{... code not shown ...}]
    }
  }
```

Matrix grid section keywords are shown in Table 6-12.

Matrix column header elements are almost identical to those shown in Table 6-13 – with the addition of the additional element keyword mapping given in Table 6-19.

Table 6-19. *Keyword Mapping for Column Headers*

Element	Keyword
Title alignment	titleAlignment

The *rowHeaders* keyword that describes the row header section uses the keywords shown in Table 6-20.

Table 6-20. *Element Keyword Mapping for the Row Header Section*

Element	Keyword
Font color	fontColor
Background color	backColor
Outline	Outline
Stepped layout	Stepped
Stepped layout indentation	steppedLayoutIndentation
URL icon	urlIcon
Word wrap	wordWrap
Font family	fontFamily
Font size	fontSize
Alignment	titleAlignment
+/– icons	showExpandCollapseButtons
Icon color	expandCollapseButtonsColor

(continued)

Table 6-20. (*continued*)

Element	Keyword
Icon size	expandCollapseButtonsSize
(Bold)	Bold
(Italic)	Italic
(Underline)	Underline

The *subTotals* keyword that describes the subtotals section uses the keywords shown in Table 6-21. There is, however, a new element that has an impact on the object nesting in the JSON theme file. In effect, an added level of nesting is applied inside the JSON that describes subtotals.

Two options are defined using the *$id* keyword in the JSON that defines the subtotals – one for column subtotals and one for row subtotals. The outcome of this is that you have *two* nearly identical subsections in the JSON that describe column and row subtotals.

Table 6-21. *Keyword Mapping for Subtotals*

Element	Keyword
Row subtotals/ColumnSubtotals	$id
Row subtotals	rowSubTotals
Column subtotals	columnSubTotals
Per column level	perColumnLevel
Per row level	perRowLevel
Column subtotals label	columnSubtotalsLabel
Row subtotals label	rowSubtotalsLabel
Row subtotal position	rowSubTotalPosition
Font color	fontColor
Font family	fontFace
Background color	backColor
(Text) size	fontSize
(Bold)	Bold
(Italic)	Italic
(Underline)	Underline
Apply to labels	applyToHeaders

Tooltip options are described in Table 6-5.

Matrix values elements are almost identical to those shown in Table 6-14 – with the addition of the additional element keyword mapping given in Table 6-22.

Table 6-22. *Keyword Mapping for Matrix Values*

Element	Keyword
Show on rows	valuesOnRow

The *rowSubTotalPosition* keyword in the JSON section for the subtotals requires one of the attributes shown in Table 6-23.

Table 6-23. *Subtotal Position Options*

Value	Formatting Popup Option
Top	Top
Bottom	Bottom

The *rowTotal* keyword that describes the Row grand total section uses the keywords shown in Table 6-24.

Table 6-24. *Keyword Mapping for Row Grand Totals*

Element	Keyword
Font	fontFamily
(Size)	fontSize
(Bold)	Bold
(Italic)	Italic
(Underline)	Underline
Apply to labels	applyToHeaders

The *columnTotal* keyword that describes the Column grand total section uses the keywords shown in Table 6-25.

Table 6-25. *Keyword Mapping for Column Grand Totals*

Element	Keyword
Font	fontFamily
(Size)	fontSize
(Bold)	Bold
(Italic)	Italic
(Underline)	Underline
Apply to labels	applyToHeaders

Table borders in the columnHeaders and rowHeaders sections are defined using the attributes of the outlineStyle keyword that are described in Table 6-17.

Conclusion

This chapter has taken you deeper into the definition of visual formatting using theme files. Specifically, you saw the complete details of how to pre-format card, multi-row card, table, and matrix visuals. You saw how they all share many common elements as well as how similar visual types can share formatting approaches. So whether you are formatting a card or a matrix, the underlying principles are the same.

The next step is to look at charts. As you will see, many chart types also share common elements. However, given the sheer number of chart types that are available in Power BI, I will be spreading this subject over three chapters. The core chart types are the subject of the next chapter.

CHAPTER 7

■ ■ ■

Classic Chart Visual Styles

Charts are a set of visuals that share many common formatting elements. So it makes sense, I feel, to look at them together. This will help your understanding of the shared aspects of these charts and also makes it easier to comprehend the more complex JSON that is used in some of the more advanced charts.

The charts that I want to look at in this chapter are

- Stacked bar chart

- Stacked column chart

- Clustered bar chart

- Clustered column chart

- Line chart

- Area chart

This is quite a collection of visuals to learn about. Fortunately, however, the clear majority of their attributes are common to most (if not all) of these chart types. This means that once you have learned the basic ways that formatting attributes are applied to the initial charts, you can apply the same code and the same knowledge to most of them. The added advantage of these core similarities is that it allows us to focus only on the differences between chart types in the JSON snippets that you will examine later in this chapter.

Each chart type discussed in this chapter has a separate JSON file in the samples folder that contains the code that you can use to customize your own theme files.

Stacked Bar Chart

The first point of note is that although the Power BI Desktop interface calls this a stacked bar chart (if you hover the pointer over the first icon in the Visualizations pane, this is what the popup calls it) in the theme file, it is called a *barChart*. This means that the keyword to remember is *barChart* as this is the keyword that introduces the JSON object that specifies stacked bar chart formatting.

Table 7-1 outlines the JSON keywords that you can use in stacked bar chart formatting and the section that corresponds to each keyword in the Power BI Desktop format pane.

© Adam Aspin 2023
A. Aspin, *Pro Power BI Theme Creation*, https://doi.org/10.1007/978-1-4842-9633-2_7

Table 7-1. *Stacked Bar Chart Keywords*

Keyword	Formatting Section
legend	Legend
categoryAxis	Y Axis
valueAxis	X Axis
zoom	Zoom slider
smallMultiplesLayout	Small Multiples
subheader	Small Multiples/Title
labels	Data labels
totals	Total labels
plotArea	Plot area
padding	Properties/Padding
title	Title/Title
subTitle	Title/Subtitle
divider	Title/Divider
spacing	Title/Spacing
Background	Effects/Background
Border	Effects/Visual border
dropShadow	Effects/Shadow
visualHeader	Header Icons
lockAspect	Lock aspect
visualHeaderTooltip	Header Icons/Help tooltip
visualTooltip	Tooltips

The sample file Attributes_StackedBarChart.json contains the complete specification of stacked bar chart JSON formatting. You can see the formatting attributes that are specific to this type of chart in the following code snippet:

```
"barChart":
  {"*":
    {
      "categoryAxis": [{
            "show": true,
            "position": "Left",
            "color": {"solid": {"color": "#949494"}},
            "fontSize": 8,
            "fontFamily": "Arial",
            "bold": true,
```

```
        "italic": false,
        "underline": false,
        "preferredCategoryWidth": 20,
        "maxMarginFactor": 25,
        "concatenateLabels": true,
        "innerPadding": 15,
        "showAxisTitle": true,
        "axisStyle": "showTitleOnly",
        "titleColor": {"solid": {"color": "#949494"}},
        "titleText": "Add an Axis Title",
        "titleFontSize": 8,
        "titleFontFamily": "Arial",
        "titleBold": true,
        "titleItalic": false,
        "titleUnderline": false,
        "switchAxisPosition": true,
        "concatenateLabels": true,
        "switchAxisPosition": false
        }],
    "valueAxis": [{
        "show": true,
        "axisScale": "linear",
        "start": "Auto",
        "end": "Auto",
        "labelColor": {"solid": {"color": "#949494"}},
        "fontSize": 8,
        "fontFamily": "Arial",
        "bold": true,
        "italic": false,
        "underline": false,
        "labelDisplayUnits": 0,
        "labelPrecision": 0,
        "labelDisplayUnits": 0,
        "showAxisTitle": true,
        "axisStyle": "showTitleOnly",
        "color": {"solid": {"color": "#949494"}},
        "titleText": "Add an Axis Title",
        "titleFontSize": 8,
        "titleFontFamily": "Arial",
        "titleBold": true,
        "titleItalic": false,
        "titleUnderline": false,
        "gridLineShow": false,
        "gridLineColor":  {"solid": {"color": "#949494"}},
        "gridLineThickness": 1,
        "gridlineStyle": "solid",
        "invertAxis": true
        }],
    "legend": [{
        "show": true,
        "position": "Top",
```

```
        "showTitle": true,
        "titleText": "Add a legend title",
        "legendColor": {"solid": {"color": "#949494"}},
        "fontFamily": "Arial",
        "fontSize": 8
        }],
"smallMultiplesLayout":   [{
        "rowCount": 3,
        "columnCount": 3,
        "gridPadding": 12,
        "advancedPaddingOptions": true,
        "columnPaddingOuter": 15,
        "columnPaddingInner": 15,
        "rowPaddingOuter": 15,
        "rowPaddingInner": 15,
        "gridLineType": "all",
        "gridLineStyle": "dotted",
        "gridLineWidth": 1,
        "gridLineColor": {"solid": {"color": "#949494"}},
        "backgroundColor": {"solid": {"color": "#949494"}},
        "backgroundTransparency": 20
        }],
"subheader":  [{
        "position": "bottom",
        "color": {"solid": {"color": "#949494"}},
        "fontSize": 8,
        "fontFamily": "Arial",
        "bold": true,
        "italic": false,
        "underline": false,
        "alignment": "center" ,
          "textWrap":  true
        }],
"zoom":[{
        "show":true,
        "showOnValueAxis": false,
        "showLabels":true,
        "showTooltip":true
        }],
"labels": [{
        "show": true,
        "color": {"solid": {"color": "#ffffff"}},
        "labelDisplayUnits": 1,
        "labelPrecision": 2,
        "labelPosition": "InsideEnd",
        "labelOverflow": true,
        "fontSize": 7,
        "fontFamily": "Arial",
        "bold": true,
        "italic": false,
        "underline": false,
```

```
              "enableBackground": false,
              "backgroundColor": {"solid": {"color": "#949494"}},
              "backgroundTransparency": 0
              }],
      "totals": [{
              "show": true,
              "color": {"solid": {"color": "#FFFFFF"}},
              "labelDisplayUnits": 1000,
              "labelPrecision": 2,
              "fontSize": 8,
              "fontFamily": "Arial",
              "bold": true,
              "italic": false,
              "underline": false,
              "enableBackground": true,
              "backgroundColor": {"solid": {"color": "#949494"}},
              "backgroundTransparency": 10,
              "showPositiveAndNegative": true
              }],
      "plotArea": [{
              "transparency": 0,
              "scaling": "Fill"
              }],
   "title": [{... code not shown ...}],
   "subTitle": [{... code not shown ...}],
   "divider": [{... code not shown ...}],
   "spacing": [{... code not shown ...}],
   "background": [{... code not shown ...}],
   "lockAspect": [{... code not shown ...}],
   "border": [{... code not shown ...}],
   "dropShadow": [{... code not shown ...}],
   "visualHeader": [{... code not shown ...}]
   "visualHeaderTooltip": [{... code not shown ...}]
   "padding": [{... code not shown ...}]
   "visualTooltip: [{... code not shown ...}]
   }
}
```

This is quite a lot of JSON code, but it does conform to the structure that you saw in the previous two chapters. In any case, I have made the JSON follow the ordering of elements in the Power BI Desktop format pane to allow for easier correlation between the JSON specification and what you see in Power BI Desktop.

There are a few comments that I need to make here:

- The vertical (Y) axis in this chart is the category axis (so the horizontal (X) axis is the value axis) in the JSON.

- The start and end values for the value axis that are set using the *start* and *end* keywords in the JSON definition can be set to specific figures, instead of null values. Simply enter the required values (as numbers without any formatting such as comma separators, etc.). If you want to leave the value empty, then you must use the *null* keyword – or the JSON theme file will fail to load.

- You cannot specify which color applies to which data point, as this is specific to the data in each different visual that you create.

- You only see the Legend card in the format pane of Power BI Desktop *if* data has been added to the Legend well in Power BI Desktop. You can, nonetheless, add the JSON that formats a legend irrespective of whether a legend is present or not as the formatting will only be applied if the legend is used in the visual.

- The axis style (set using the axisStyle keyword) is limited to Show title only in stacked bar (and column) charts.

The *legend* keyword that describes the legend section uses the keywords shown in Table 7-2.

Table 7-2. *Keyword Mapping for Legends*

Element	Keyword
Legend	Show
Position	Position
Title	showTitle
Legend name	titleText
Color	legendColor
Font family	fontFamily
Text size	fontSize

The *categoryAxis* keyword that describes the Y axis section uses the keywords shown in Table 7-3.

Table 7-3. *Keyword Mapping for the Y Axis*

Element	Keyword
X axis	Show
Position	Position
Color	Color
Text size	fontSize
Font family	fontFamily
(Bold)	Bold
(Italic)	Italic
(Underline)	Underline
Minimum category width	preferredCategoryWidth
Maximum size	maxMarginFactor
Concatenate Labels	concatenateLabels
Inner padding	innerPadding
Title	showAxisTitle
Style	axisStyle
Title color	titleColor
Axis title	titleText
Title text size	titleFontSize
(Title) Font family	titleFontFamily
(Title) Bold	titleBold
(Title) Italic	titleItalic
(Title) Underline	titleUnderline
Switch axis position	switchAxisPosition
Concatenate labels	concatenateLabels
Switch axis position	switchAxisPosition

The *valueAxis* keyword that describes the X axis section uses the keywords shown in Table 7-4.

Table 7-4. *Keyword Mapping for the X Axis*

Element	Keyword
Y axis	Show
Scale type	axisScale
Start	Start
End	End
Color	labelColor
Text size	fontSize
Font family	fontFamily
(Bold)	Bold
(Italic)	Italic
(Underline)	Underline
Display units	labelDisplayUnits
Value decimal places	labelPrecision
Title	showAxisTitle
Style	axisStyle
Title color	Color
Axis title	titleText
Title text size	titleFontSize
(Title) Font family	titleFontFamily
(Title) Bold	titleBold
(Title) Italic	titleItalic
(Title) Underline	titleUnderline
Gridlines	gridLineShow
(Gridline) Color	gridLineColor
Stroke width	gridLineThickness
Line style	gridlineStyle
Invert range	invertAxis

The *zoom* keyword that describes the zoom slider section uses the keywords shown in Table 7-5.

Table 7-5. *Keyword Mapping for the Zoom Slider*

Element	Keyword
Zoom slider	Show
X axis	showOnValueAxis
Slider labels	showLabels
Slider tooltips	showTooltip

The *labels* keyword that describes the data labels section uses the keywords shown in Table 7-6.

Table 7-6. *Keyword Mapping for Data Labels*

Element	Keyword
Data labels	Show
Color	Color
Display units	labelDisplayUnits
Value decimal places	labelPrecision
Position	labelPosition
Overflow text	labelOverflow
Text size	fontSize
Font family	fontFamily
(Bold)	Bold
(Italic)	Italic
(Underline)	Underline
Show background	enableBackground
Background color	backgroundColor
Transparency	backgroundTransparency

The *totals* keyword that describes the total labels section uses the keywords shown in Table 7-7.

Table 7-7. *Keyword Mapping for Total Labels*

Element	Keyword
Total labels	Show
Color	Color
Display units	labelDisplayUnits
Value decimal places	labelPrecision
Text size	fontSize
Font family	fontFamily
(Bold)	Bold
(Italic)	Italic
(Underline)	Underline
Show background	enableBackground
Background color	backgroundColor
Transparency	backgroundTransparency
Split positive and negative	showPositiveAndNegative

The *plotArea* keyword that describes the plot area section uses the keywords shown in Table 7-8.

Table 7-8. *Keyword Mapping for the Plot Area*

Element	Keyword
Plot area	Show
Transparency	Transparency
Scaling	Scaling

Tooltip options are described in Table 6-5.

The *smallMultiplesLayout* keyword that describes the Small multiples section uses the keywords shown in Table 7-9.

Table 7-9. *Keyword Mapping for Small Multiples*

Element	Keyword
Layout/Rows	rowCount
Layout/Columns	columnCount
Layout/All padding	gridPadding
Layout/Customize padding	advancedPaddingOptions
Layout/Outer column padding	columnPaddingOuter
Layout/Inner column padding	columnPaddingInner
Layout/Outer row padding	rowPaddingOuter
Layout/Inner row padding	rowPaddingInner
Border/Gridlines	gridLineType
Border/Line style	gridLineStyle
Border/Line color	gridLineColor
Border/Line width	gridLineWidth
Background/Color	backgroundColor
Background/Transparency	backgroundTransparency

The *subtitle* keyword that describes the Small multiples title section uses the keywords shown in Table 7-10.

Table 7-10. *Keyword Mapping for Small Multiples Titles*

Element	Keyword
Title/Show	Show
Title/Position	Position
Title/Font	fontFamily
Title/Font color	Color
Title/(Size)	fontSize
Title/(Bold)	Bold
Title/(Italic)	Italic
Title/(Underline)	Underline
Title/(Alignment)	Alignment
Title/Text wrap	textWrap

The *position* keyword in the JSON that is used in the legend section for the legend position inside a chart needs one of the attributes shown in Table 7-11.

Table 7-11. *Legend Position Options*

Value	Formatting Popup Option
TopLeft	Top left
Bottom	Bottom
Left	Top left stacked
Right	Top right stacked
TopCenter	Top center
TopRight	Top right
BottomCenter	Bottom center
BottomLeft	Bottom left
BottomRight	Bottom right
LeftCenter	Center left
RightCenter	Center right

■ **Note** As these are *attributes* (values in JSON) and not *keywords*, they do not necessarily conform to the camel case requirement that applies to "pure" keywords. Also, any text values *must* be enclosed in double quotes.

The *position* keyword in the JSON in the category axis section needs one of the attributes shown in Table 7-12.

Table 7-12. *Category Axis Position Options*

Value	Formatting Popup Option
Left	Left
Right	Right

The *labelPrecision* keyword in the JSON that is used in the label section of the values axis needs one of the attributes shown in Table 7-13.

Table 7-13. *Label Display Units Options*

Value	Formatting Popup Option
0	Auto
1	None
1000	Thousands
1000000	Millions
1000000000	Billions
1000000000000	Trillions

■ **Note** As these are *numeric* values in JSON, they must *not* be enclosed in double quotes.

The *labelPosition* keyword in the JSON that is used in the label section needs one of the attributes shown in Table 7-14.

Table 7-14. *Label Position Options*

Value	Formatting Popup Option
Auto	Auto
InsideEnd	Inside end
InsideCenter	Inside center
InsideBase	Inside base

The *gridLineType* keyword in the JSON that is used in the Small multiples border needs one of the attributes shown in Table 7-15.

Table 7-15. *Gridline Options*

Value	Formatting Popup Option
all	All
none	None
inner	Horizontal and vertical
innerHorizontal	Horizontal only
innerVertical	Vertical only

The *gridLineStyle* keyword in the JSON that is used in the X axis and the small multiples border needs one of the attributes shown in Table 7-16.

Table 7-16. *Value Axis Gridline Style Options*

Value	Formatting Popup Option
Dotted	Dotted
Dashed	Dashed
Solid	Solid

■ **Note** The customize series option (for data labels) seems, as with all formatting that requires a source field to be selected, not to be configurable using theme files.

Stacked Column Chart

The next chart to look at is the stacked column chart. The JSON is largely identical to the code you saw earlier, which makes the point that Power BI charts share similar structural approaches.

The keyword that specifies stacked column chart formatting attributes is *columnChart*.

Table 7-17 outlines the JSON keywords that you can use in stacked bar chart formatting and the section that corresponds to each keyword in the Power BI Desktop format pane.

Table 7-17. *Stacked Bar Chart Keywords*

Keyword	Formatting Section
legend	Legend
categoryAxis	X Axis
valueAxis	Y Axis
zoom	Zoom slider
smallMultiplesLayout	Small Multiples
subheader	Small Multiples/Title
labels	Data labels
Totals	Total labels
plotArea	Plot area
padding	Properties/Padding
Title	Title/Title
subTitle	Title/Subtitle
divider	Title/Divider
spacing	Title/Spacing
background	Effects/Background
border	Effects/Visual border
dropShadow	Effects/Shadow
visualHeader	Header Icons
lockAspect	Lock aspect
visualHeaderTooltip	Header Icons
visualTooltip	Tooltips

The fundamental differences with the JSON for the stacked bar chart that you saw previously are

- The *categoryAxis* has the *concatenateLabels* keyword added – and the *position* keyword removed.

- The *valueAxis* has the *position* keyword added – and the *axisScale* keyword removed.

- Labels has labelOrientation added.

- There are now also options to choose the category and value axis style that are available for this chart type.

- The vertical (Y) axis in this chart is the value axis (so the horizontal (X) axis is the category axis) in the JSON.

The complete definition is available in the file Attributes_StackedColumnChart.json in the sample files. The JSON code is

```
"columnChart":
 {"*":
   {
   "legend": [{... code not shown ...}],
    "categoryAxis": [{
          "show": true,
          "color": {"solid": {"color": "#949494"}},
          "fontSize": 8,
          "fontFamily": "Arial",
          "bold": false,
          "italic": false,
          "underline": false,
          "preferredCategoryWidth": 20,
          "maxMarginFactor": 25,
          "innerPadding": 15,
          "concatenateLabels": true,
          "showAxisTitle": true,
          "axisStyle": "showTitleOnly",
          "titleColor": {"solid": {"color": "#949494"}},
          "titleText": "Add a category Axis Title",
          "titleFontSize": 8,
          "titleFontFamily": "Arial",
          "bold": false,
          "italic": false,
          "underline": false,
          "concatenateLabels": true
          }],
     "valueAxis": [{
          "show": true,
          "position": "Left",
          "start": "Auto",
          "end": "Auto",
          "labelColor": {"solid": {"color": "#949494"}},
          "fontSize": 8,
          "fontFamily": "Arial",
          "bold": false,
          "italic": false,
          "underline": false,
          "labelPrecision": 0,
          "labelDisplayUnits": 0,
          "invertAxis": false,
          "showAxisTitle": true,
          "axisStyle": "showUnitOnly",
          "color": {"solid": {"color": "#949494"}},
          "titleText": "Add a Value Axis Title",
          "titleFontSize": 8,
          "titleFontFamily": "Arial",
          "titleBold": true,
```

```
            "titleItalic": false,
            "titleUnderline": false,
            "gridLineShow": false,
            "gridLineColor":  {"solid": {"color": "#949494"}},
            "gridLineThickness": 1,
            "gridlineStyle": "solid",
            "invertAxis": true
        }],
    "labels": [{
            "show": true,
            "color": {"solid": {"color": "#ffffff"}},
            "labelDisplayUnits": 1,
            "labelPrecision": 2,
            "labelOrientation": 1,
            "labelPosition": "InsideEnd",
            "labelOverflow": true,
            "fontFamily": "Arial",
            "fontSize": 7,
            "enableBackground": false,
            "backgroundColor": {"solid": {"color": "#949494"}},
            "backgroundTransparency": 0
        }],
    "zoom: [{... code not shown ...}],
    "smallMultiplesLayout: [{... code not shown ...}],
    "subHeader: [{... code not shown ...}],
    "totals": [{... code not shown ...}],
    "plotArea": [{... code not shown ...}],
    "title": [{... code not shown ...}],
    "subTitle": [{... code not shown ...}],
    "divider": [{... code not shown ...}],
    "spacing": [{... code not shown ...}],
    "background": [{... code not shown ...}],
    "lockAspect": [{... code not shown ...}],
    "border": [{... code not shown ...}],
    "dropShadow": [{... code not shown ...}],
    "visualHeader": [{... code not shown ...}]
    "visualHeaderTooltip": [{... code not shown ...}]
    "padding": [{... code not shown ...}]
    "visualTooltip: [{... code not shown ...}]
    }
}
```

The stacked column chart keywords for the X Axis, Y axis, and labels sections are, fortunately, largely identical to those used for stacked bars – so I will not repeat them here, but prefer to refer you to the previous section instead.

The *labelOrientation* keyword in the JSON that sets label orientation (in the labels section) needs one of the attributes shown in Table 7-18 to be supplied.

Table 7-18. *Label Display Units Options*

Value	Formatting Popup Option
Vertical	Vertical
Horizontal	Horizontal

The *axisStyle* keyword in the JSON for the value axis needs one of the attributes shown in Table 7-19.

Table 7-19. *Category and Value Axis Style Options*

Value	Formatting Popup Option
ShowTitleOnly	Show title only
ShowUnitOnly	Show unit only
ShowBoth	Show both

Legend keywords are given in Table 7-2.
Categories keywords are given in Table 7-3.
Values keywords are given in Table 7-4.
Zoom keywords are given in Table 7-5. For a column chart, the showOnValueAxis refers to the Y axis.
Labels keywords are given in Table 7-6.
Totals keywords are given in Table 7-7.
Plot area keywords are given in Table 7-8.
Tooltip keywords are given in Table 6-5.
Small multiples options are described in Table 7-9.
Small multiples titles options are described in Table 7-10.
Legend position options are described in Table 7-11.
Category axis position options are described in Table 7-12.
Value axis display units options are described in Table 7-13.
Label position options are described in Table 7-14.
Gridline options are described in Table 7-15.

Clustered Bar Chart

The JSON that specifies clustered bar chart formatting is virtually identical to the code that you saw earlier for a stacked bar chart – except for the initial keyword for the JSON object, of course. The only major difference is that it does not contain a *totals* section.

The keyword that specifies clustered bar chart formatting attributes is *clusteredBarChart*.

Table 7-20 outlines the JSON keywords that you can use in clustered bar chart formatting and the section that corresponds to each keyword in the Power BI Desktop format pane.

Table 7-20. *Clustered Bar Chart Keywords*

Keyword	Formatting Section
Legend	Legend
categoryAxis	Y Axis
valueAxis	X Axis
Zoom	Zoom slider
smallMultiplesLayout	Small Multiples
Subheader	Small Multiples/Title
Labels	Data labels
plotArea	Plot area
Padding	Properties/Padding
Title	Title/Title
subtitle	Title/Subtitle
Divider	Title/Divider
Spacing	Title/Spacing
Background	Effects/Background
Border	Effects/Visual border
dropShadow	Effects/Shadow
visualHeader	Header Icons/Icons Header Icons/Colors
lockAspect	Lock aspect
visualHeaderTooltip	Header Icons/Help tooltip
visualTooltip	Tooltips

The JSON code that specifies formatting for a clustered bar chart is available in the sample file Attributes_ClusteredBarChart.json.

In the interest of completeness, here is the JSON code that defines the formatting of a clustered bar chart:

```
"clusteredBarChart":
  {"*" :
    {
    "legend": [{... code not shown ...}],
    "categoryAxis": [{
        "show": true,
        "position": "Left",
        "color": {"solid": {"color": "#949494"}},
        "fontSize": 8,
        "fontFamily": "Arial",
```

```
            "preferredCategoryWidth": 20,
            "maxMarginFactor": 25,
            "innerPadding": 15,
            "showAxisTitle": true,
            "axisStyle": "showTitleOnly",
            "titleColor": {"solid": {"color": "#949494"}},
            "titleText": "Add an Axis Title",
            "titleFontSize": 8,
            "titleFontFamily": "Arial",
            "concatenateLabels": true,
            "switchAxisPosition": false
            }],
     "valueAxis": [{
            "show": true,
            "axisScale": "Linear",
            "start": null,
            "end": null,
            "labelColor": {"solid": {"color": "#949494"}},
            "fontSize": 8,
            "fontFamily": "Arial",
            "labelPrecision": 0,
            "labelDisplayUnits": 2,
            "showAxisTitle": true,
            "axisStyle": "ShowTitleOnly",
            "color": {"solid": {"color": "#949494"}},
            "titleText": "Add an Axis Title",
            "titleFontSize": 8,
            "titleFontFamily": "Arial",
            "gridLineShow": false,
            "gridLineColor":  {"solid": {"color": "#949494"}},
            "gridLineThickness": 1,
            "gridlineStyle": "Solid",
            "invertAxis": true
         }],
      "zoom: [{... code not shown ...}]
       "smallMultiplesLayout: [{... code not shown ...}],
      "subHeader: [{... code not shown ...}],
     "labels": [{
            "show": true,
            "color": {"solid": {"color": "#FFFFFF"}},
            "labelDisplayUnits": 1,
            "labelPrecision": 2,
            "labelPosition": "InsideEnd",
            "labelOverflow": true,
            "fontFamily": "Arial",
            "fontSize": 7,
            "enableBackground": false,
            "backgroundColor": {"solid": {"color": "#949494"}},
            "backgroundTransparency": 0
            }],
       "plotArea": [{... code not shown ...}],
```

```
"title": [{... code not shown ...}],
"subTitle": [{... code not shown ...}],
"divider": [{... code not shown ...}],
"spacing": [{... code not shown ...}],
"background": [{... code not shown ...}],
"lockAspect": [{... code not shown ...}],
"border": [{... code not shown ...}],
"dropShadow": [{... code not shown ...}],
"visualHeader": [{... code not shown ...}]
"visualHeaderTooltip": [{... code not shown ...}]
"padding": [{... code not shown ...}]
"visualTooltip: [{... code not shown ...}]
 }
}
```

Legend keywords are given in Table 7-2.

Categories keywords are given in Table 7-3.

Values keywords are given in Table 7-4.

Zoom keywords are given in Table 7-5. For a column chart, the showOnValueAxis refers to the X axis.

Labels keywords are given in Table 7-6.

Totals keywords are given in Table 7-7.

Plot area keywords are given in Table 7-8.

Tooltip keywords are given in Table 6-5.

Small multiples options are described in Table 7-9.

Small multiples titles options are described in Table 7-10.

Legend position options are described in Table 7-11.

Category axis position options are described in Table 7-12.

Value axis display units options are described in Table 7-13.

Label position options are described in Table 7-14.

Gridline options are described in Table 7-15.

Clustered Column Chart

The JSON that specifies clustered column chart formatting is very similar to the code that you saw earlier for a stacked column chart – except for the initial keyword for the JSON object, of course. This is *clusteredColumnChart*.

The key differences are

- It does *not* contain a *totals* section.

- The *labelOrientation* keyword has been added to the *labels* section.

Table 7-21 outlines the JSON keywords that you can use in stacked bar chart formatting and the section that corresponds to each keyword in the Power BI Desktop format pane.

Table 7-21. *Clustered Column Chart Keywords*

Keyword	Formatting Section
legend	Legend
categoryAxis	X Axis
valueAxis	Y Axis
zoom	Zoom slider
smallMultiplesLayout	Small Multiples
subheader	Small Multiples/Title
labels	Data labels
plotArea	Plot area
title	Title
lockAspect	Lock aspect
border	Border
dropShadow	Shadow
visualHeader	Visual header
padding	Properties/Padding
title	Title/Title
subTitle	Title/Subtitle
divider	Title/Divider
spacing	Title/Spacing
background	Effects/Background
border	Effects/Visual border
dropShadow	Effects/Shadow
visualHeader	Header Icons
lockAspect	Lock aspect
visualHeaderTooltip	Header Icons/Help tooltip
visualTooltip	Tooltips

The following code snippet provides the JSON code used to pre-format a clustered column chart:

```
"clusteredColumnChart":
  {"*":
    {
    "legend": [{... code not shown ...}],
    "categoryAxis": [{
        "show": true,
```

```
            "color": {"solid": {"color": "#949494"}},
            "fontSize": 8,
            "fontFamily": "Arial",
            "preferredCategoryWidth": 20,
            "maxMarginFactor": 25,
            "innerPadding": 15,
            "concatenateLabels": true,
            "showAxisTitle": true,
            "axisStyle": "showTitleOnly",
            "titleColor": {"solid": {"color": "#949494"}},
            "titleText": "Add an Axis Title",
            "titleFontSize": 8,
            "titleFontFamily": "Arial",
            "concatenateLabels": true
        }],
    "valueAxis": [{
            "show": true,
            "position": "Left",
            "start": null,
            "end": null,
            "labelColor": {"solid": {"color": "#949494"}},
            "fontSize": 8,
            "fontFamily": "Arial",
            "labelPrecision": 0,
            "labelDisplayUnits": 0,
            "showAxisTitle": true,
            "axisStyle": "ShowUnitOnly",
            "color": {"solid": {"color": "#949494"}},
            "titleText": "Add an Axis Title",
            "titleFontSize": 8,
            "titleFontFamily": "Arial",
            "gridLineShow": false,
            "gridLineColor":  {"solid": {"color": "#949494"}},
            "gridLineThickness": 1,
            "gridlineStyle": "Solid",
            "invertAxis": true
        }],
    "zoom: [{... code not shown ...}],
    "smallMultiplesLayout: [{... code not shown ...}],
    "subHeader: [{... code not shown ...}],
    "labels": [{
            "show": true,
            "color": {"solid": {"color": "#ffffff"}},
            "labelDisplayUnits": 1,
            "labelPrecision": 2,
            "labelOrientation": 1,
            "labelPosition": "InsideEnd",
            "labelOverflow": true,
            "fontFamily": "Arial",
            "fontSize": 7,
            "enableBackground": false,
```

```
            "backgroundColor": {"solid": {"color": "#949494"}},
            "backgroundTransparency": 0
            }],
    "plotArea": [{... code not shown ...}],
    "title": [{... code not shown ...}],
    "background": [{... code not shown ...}],
    "lockAspect": [{... code not shown ...}],
    "border": [{... code not shown ...}],
    "dropShadow": [{... code not shown ...}],
    "visualTooltip": [{... code not shown ...}],
    "visualHeader": [{... code not shown ...}]
    }
}
```

The *labelOrientation* keyword in the JSON that sets label orientation (in the labels section) needs one of the attributes shown in Table 7-22 to be supplied.

Table 7-22. *Label Display Units Options*

Value	Formatting Popup Option
vertical	0
horizontal	1

■ **Note** The clustered column chart uses 0 and 1 instead of horizontal and vertical as keywords – unlike the stacked column chart. Also, these are numeric values and so must not be enclosed in double quotes.

It is worth noting that the JSON for a clustered column chart also has options (as for clustered bar chart) to choose the category and value axis style.

The JSON code for a clustered column chart is available in the sample file Attributes_ClusteredColumnChart.json.

Legend keywords are given in Table 7-2.

Categories keywords are given in Table 7-3.

Values keywords are given in Table 7-4.

Zoom keywords are given in Table 7-5. For a column chart, the showOnValueAxis refers to the Y axis.

Labels keywords are given in Table 7-6.

Totals keywords are given in Table 7-7.

Plot area keywords are given in Table 7-8.

Tooltip keywords are given in Table 6-5.

Small multiples options are described in Table 7-9.

Small multiples titles options are described in Table 7-10.

Legend position options are described in Table 7-11.

Category axis position options are described in Table 7-12.

Value axis display units options are described in Table 7-13.

Label position options are described in Table 7-14.

Gridline options are described in Table 7-15.

Line Chart

Although they look very different to bar and column charts, the formatting aspects of line charts are really quite similar to all the chart types that you have seen so far in this chapter.

The key differences that you will find in the JSON for a line chart – as compared to bar and column charts – are

- The line chart contains JSON to describe a secondary axis.

- The legend contains a few additional elements.

- There are slight differences in the X (category) axis.

- The minimum category width still uses the keyword *preferredCategoryWidth*.

- The maximum size still uses the keyword *maxMarginFactor*.

- There is an additional element to describe line styles – introduced by the *lineStyles* keyword.

- The *labels* section contains a *labelPosition* element – but the values are different to those you have seen for other charts.

The keyword that specifies the formatting attributes of a line chart is *lineChart*.

Table 7-23 outlines the JSON keywords that you can use in line chart formatting and the section that corresponds to each keyword in the Power BI Desktop format pane.

Table 7-23. *Line Chart Keywords*

Keyword	Formatting Section
legend	Legend
categoryAxis	X Axis
valueAxis	Y Axis
zoom	Zoom slider
smallMultiplesLayout	Small Multiples
subheader	Small Multiples/Title
labels	Data labels
lineStyles	Shapes
seriesLabels	Series labels
plotArea	Plot area
padding	Properties/Padding
title	Title/Title
subTitle	Title/Subtitle
divider	Title/Divider
spacing	Title/Spacing
background	Effects/Background
border	Effects/Visual border
dropShadow	Effects/Shadow
visualHeader	Header Icons/Icons Header Icons/Colors
lockAspect	Lock aspect
visualHeaderTooltip	Header Icons/Help tooltip
visualTooltip	Tooltips

The JSON code to describe the formatting attributes of a line chart is in the sample file Attributes_ LineChart.json. The code snippet is given as follows. It is, be warned, a fairly long piece of JSON.

```
"lineChart":
  {"*":
    {
      "categoryAxis": [{
            "show": true,
            "axisType":"Categorical",
            "color": {"solid": {"color": "#949494"}},
            "fontSize": 8,
            "fontFamily": "Arial",
```

```
        "bold": true,
        "italic": false,
        "underline": false,
        "preferredCategoryWidth": 20,
        "maxMarginFactor": 30,
        "concatenateLabels": true,
        "showAxisTitle": true,
        "axisStyle": "showTitleOnly",
        "titleColor": {"solid": {"color": "#949494"}},
        "titleText": "Add an Axis Title",
        "titleFontSize": 8,
        "titleFontFamily": "Arial",
        "titleBold": true,
        "titleItalic": false,
        "titleUnderline": false,
        "concatenateLabels": true
        }],
"valueAxis": [{
        "show": true,
        "position": "Right",
        "axisScale": "linear",
        "start": "Auto",
        "end": "Auto",
        "labelColor": {"solid": {"color": "#949494"}},
        "fontSize": 8,
        "fontFamily": "Arial",
        "bold": true,
        "italic": false,
        "underline": false,
        "labelPrecision": 0,
        "labelDisplayUnits": 0,
        "showAxisTitle": true,
        "axisStyle": "showUnitOnly",
        "color": {"solid": {"color": "#949494"}},
        "titleText": "Add an Axis Title",
        "titleFontSize": 8,
        "titleFontFamily": "Arial",
        "titleBold": true,
        "titleItalic": false,
        "titleUnderline": false,
        "gridLineShow": false,
        "gridLineColor":  {"solid": {"color": "#949494"}},
        "gridLineThickness": 2,
        "gridlineStyle": "dotted",
        "invertAxis": true
      }],
"y2Axis": [{
        "show": true,
        "secStart": 0,
        "secEnd": 0,
        "secAxisScale": "linear",
```

```
        "secLabelColor": {"solid": {"color": "#949494"}},
        "secFontSize": 8,
        "secSontFamily": "Arial",
        "secBold": true,
        "secItalic": false,
        "secUnderline": false,
        "secLabelPrecision": 0,
        "secLabelDisplayUnits": 0,
        "secShowAxisTitle": true,
        "secAxisStyle": "showUnitOnly",
        "secColor": {"solid": {"color": "#949494"}},
        "secTitleText": "Add an Axis Title",
        "secTitleFontSize": 8,
        "secTitleFontFamily": "Arial",
        "secTitleBold": true,
        "secTitleItalic": false,
        "secTitleUnderline": false
        }],
    "legend": [{
        "show": true,
        "position": "Top",
        "showTitle": true,
        "titleText": "Add a legend title",
        "legendColor": {"solid": {"color": "#949494"}},
        "fontFamily": "Arial",
        "fontSize": 8,
        "bold": true,
        "italic": false,
        "underline": false,
        "legendMarkerRendering":"lineOnly",
        "matchLineColor":true,
        "defaultToCircle":true
        }],
    "labels": [{
        "show": true,
        "color": {"solid": {"color": "#ffffff"}},
        "labelDisplayUnits": 1,
        "labelPrecision": 2,
        "labelPosition": "Above",
        "labelOverflow": true,
        "fontFamily": "Arial",
        "fontSize": 7,
        "enableBackground": false,
        "backgroundColor": {"solid": {"color": "#949494"}},
        "backgroundTransparency": 0
        }],
    "seriesLabels": [{
        "show": true,
        "seriesMaximumWidth": 1,
        "seriesPosition": "Left",
        "seriesFontFamily": "Arial",
```

```
            "textSize": 9,
            "bold": false,
            "italic": true,
            "underline": false,
            "seriesColor": {"solid": {"color": "#949494"}},
            "seriesWordWrap": false,
            "enableBackground": false,
            "backgroundColor": {"solid": {"color": "#12239E"}},
            "backgroundTransparency": 0
        }],
    "lineStyles": [{
            "strokeWidth": 1,
            "strokeLineJoin": "bevel",
            "lineStyle": "solid",
            "showMarker": true,
            "markerShape": "longDash",
            "markerSize": 3,
            "markerColor":  {"solid": {"color": "#949494"}},
            "stepped": true,
            "showSeries": false
        }],
    "zoom: [{... code not shown ...}],
     "smallMultiplesLayout: [{... code not shown ...}],
    "subHeader: [{... code not shown ...}],
    "plotArea": [{... code not shown ...}],
    "title": [{... code not shown ...}],
    "subTitle": [{... code not shown ...}],
    "divider": [{... code not shown ...}],
    "spacing": [{... code not shown ...}],
    "background": [{... code not shown ...}],
    "lockAspect": [{... code not shown ...}],
    "border": [{... code not shown ...}],
    "dropShadow": [{... code not shown ...}],
    "visualHeader": [{... code not shown ...}]
    "visualHeaderTooltip": [{... code not shown ...}]
    "padding": [{... code not shown ...}]
    "visualTooltip: [{... code not shown ...}]
    }
}
```

As well as the elements in the new line styles and secondary Y axis sections, line charts have an option that you have not seen in chart visuals so far. This is the axis type (in the category axis). In fact, the legend and category axis sections are sufficiently different to merit specific review.

The *legend* keyword that describes the legend section uses the keywords shown in Table 7-24.

Table 7-24. *Keyword Mapping for Legend*

Element	Keyword
Legend	Show
Position	Position
Title	showTitle
Legend name	titleText
Color	legendColor
Font family	fontFamily
Text size	fontSize
(Bold)	Bold
(Italic)	Italic
(Underline)	Underline
Style	legendMarkerRendering
Match line color	matchLineColor
Circle default icon	defaultToCircle

The *categoryAxis* keyword that describes the X axis section uses the keywords shown in Table 7-25.

Table 7-25. *Keyword Mapping for the X Axis*

Element	Keyword
X axis	Show
Axis type	axisType
Color	Color
Text size	fontSize
Font family	fontFamily
(Bold)	Bold
(Italic)	Italic
(Underline)	Underline
Minimum category width	preferredCategoryWidth
Maximum size	maxMarginFactor
Title	showAxisTitle
Style	axisStyle
Title color	titleColor
Axis title	titleText
Title text size	titleFontSize
(Title) Font family	titleFontFamily

As always, the value that you use in the JSON theme must match – exactly – one of the expected values, or the default value will remain active. Equally, remember that the values are case-sensitive.

The *valueAxis* options are identical to those shown in Table 7-3, with the addition of the *position* element.

The *y2Axis* keyword that describes the secondary Y axis section uses the keywords shown in Table 7-26.

Table 7-26. *Keyword Mapping for the Secondary Y Axis*

Element	Keyword
Secondary Y axis	secShow
Start	secStart
End	secEnd
Scale type	secAxisScale
Color	secLabelColor
Text size	secFontSize
Font family	secFontFamily
(Bold)	secBold
(Italic)	secItalic
(Underline)	secUnderline
Display units	secLabelDisplayUnits
Value decimal places	secLabelPrecision
Title	secShowAxisTitle
Style	secAxisStyle
Title color	secColor
Axis title	secTitleText
Title text size	secTitleFontSize
(Title) Font family	secTitleFontFamily
(Title) Bold	secTitleBold
(Title) Italic	secTitleItalic
(Title) Underline	secTitleUnderline
Gridlines	secGridLineShow

Note that for the secondary axis, the only values that are accepted are 0. This means that the keyword "auto" cannot be entered as is the case for the main Y axis. Hopefully, this will be corrected by the time that this book is in print.

Zoom options are shown in Table 7-5.

Label options are shown in Table 7-6.

The *seriesLabels* keyword that describes the Series labels section uses the keywords shown in Table 7-27.

Table 7-27. *Keyword Mapping for the Series Labels*

Element	Keyword
Series labels	Show
Maximum width	seriesMaximumWidth
Series position	seriesPosition
Font	seriesFontFamily
(Size)	textSize
(Bold)	Bold
(Italic)	Italic
(Underline)	Underline
Background	enableBackground
Background color	backgroundColor
Transparency	backgroundTransparency

The *labelPosition* keyword that is used in the *labels* section needs one of the attributes shown in Table 7-28.

Table 7-28. *Label Position Options*

Element	Keyword
None	None
Above	Above
okUnder	Under

The *lineStyle* keyword that describes the Lines section uses the keywords shown in Table 7-29.

Table 7-29. *Keyword Mapping for Line Styles*

Element	Keyword
Stroke width	strokeWidth
Join type	strokeLineJoin
Line style	lineStyle
Show marker	showMarker
Marker shape	markerShape
Marker size	markerSize
Marker color	markerColor
Stepped	Stepped
Customize series	showSeries

The *axisType* keyword in the JSON that defines the category axis in a line (or area or stacked area) chart needs one of the attributes shown in Table 7-30.

Table 7-30. *X Axis Type Options*

Value	Formatting Popup Option
Categorical	Categorical
Continuous	Continuous

The *axisScale* keyword in the JSON that specifies the category axis scale needs one of the attributes shown in Table 7-31.

Table 7-31. *Value Axis Scale Options*

Value	Formatting Popup Option
Linear	Linear
Log	Log

The *lineStyle* keyword in the JSON for the line style section needs one of the attributes shown in Table 7-32.

Table 7-32. *Line Style Options*

Value	Formatting Popup Option
markerOnly	Markers only
lineAndMarker	Line and Markers
lineOnly	Line only

The *strokeLineJoin* keyword in the JSON that defines the shape of the join in the line style section needs one of the attributes shown in Table 7-33.

Table 7-33. *Join Type Options*

Value	Formatting Popup Option
Miter	Miter
Round	Round
Bevel	Bevel

The *markerShape* keyword in the JSON that defines the shape of the marker in the line style section needs one of the attributes shown in Table 7-34.

Table 7-34. *Marker Shape Options*

Value	Formatting Popup Option
Round	Round
Square	Square
Diamond	Diamond
Triangle	Triangle
x	X
shortDash	-
longDash	—
+	+

Small multiples options are described in Table 7-9.
Small multiples titles options are described in Table 7-10.
Legend position options are described in Table 7-11.
Category axis position options are described in Table 7-12.
Value axis display units options are described in Table 7-13.
Category axis display units options are described in Table 7-13.
Label position options are described in Table 7-14.
Gridline options are described in Table 7-15.
Value axis gridline style options are described in Table 7-16.
As far as I can make out, it is not currently possible to define the formatting of customized series (shapes/customize series in the Power BI Desktop format pane). I can understand why this would be the case, as this option needs a field to be specified – and so cannot be defined generically.

Area Chart

The JSON used to specify how an area chart is formatted is nearly identical to the code used for a line chart, apart from using a different initial keyword (i.e., you have to use *areaChart*).

Table 7-35 outlines the JSON keywords that you can use in area chart formatting and the section that corresponds to each keyword in the Power BI Desktop format pane.

Table 7-35. *Area Chart Keywords*

Keyword	Formatting Section
legend	Legend
categoryAxis	X Axis
valueAxis	Y Axis
zoom	Zoom slider
smallMultiplesLayout	Small Multiples
subheader	Small Multiples/Title
labels	Data labels
lineStyles	Shapes
plotArea	Plot area
padding	Properties/Padding
title	Title/Title
subTitle	Title/Subtitle
divider	Title/Divider
spacing	Title/Spacing
background	Effects/Background
border	Effects/Visual border
dropShadow	Effects/Shadow
visualHeader	Header Icons/Icons Header Icons/Colors
lockAspect	Lock aspect
visualHeaderTooltip	Header Icons/Help tooltip
visualTooltip	Tooltips

The full JSON code to describe the formatting attributes of an area chart is available in Attributes_AreaChart.json in the sample files.

The following is the JSON that describes the formatting of an area chart:

```
"areaChart":
  {"*" :
    {
      "categoryAxis": [{
            "show": true,
            "axisType":"Categorical",
            "color": {"solid": {"color": "#949494"}},
            "fontSize": 8,
            "fontFamily": "Arial",
            "bold": true,
```

```
        "italic": false,
        "underline": false,
        "preferredCategoryWidth": 20,
        "maxMarginFactor": 30,
        "concatenateLabels": true,
        "showAxisTitle": true,
        "axisStyle": "showTitleOnly",
        "titleColor": {"solid": {"color": "#949494"}},
        "titleText": "Add an Axis Title",
        "titleFontSize": 8,
        "titleFontFamily": "Arial",
        "titleBold": true,
        "titleItalic": false,
        "titleUnderline": false,
        "concatenateLabels": true
        }],
"valueAxis": [{
        "show": true,
        "position": "Left",
        "axisScale": "linear",
        "start": "Auto",
        "end": "Auto",
        "labelColor": {"solid": {"color": "#949494"}},
        "fontSize": 8,
        "fontFamily": "Arial",
        "bold": true,
        "italic": false,
        "underline": false,
        "labelPrecision": 0,
        "labelDisplayUnits": 0,
        "showAxisTitle": true,
        "axisStyle": "showUnitOnly",
        "color": {"solid": {"color": "#949494"}},
        "titleText": "Add an Axis Title",
        "titleFontSize": 8,
        "titleFontFamily": "Arial",
        "titleBold": true,
        "titleItalic": false,
        "titleUnderline": false,
        "gridLineShow": true,
        "gridLineColor":  {"solid": {"color": "#949494"}},
        "gridLineThickness": 2,
        "gridlineStyle": "dashed",
        "invertAxis": true
      }],
"y2Axis": [{
        "show": true,
        "secStart": 0,
        "secEnd": 0,
        "secAxisScale": "linear",
        "secLabelColor": {"solid": {"color": "#949494"}},
```

```
              "secFontSize": 8,
              "secFontFamily": "Arial",
              "secBold": true,
              "secItalic": false,
              "secUnderline": false,
              "secLabelPrecision": 0,
              "secLabelDisplayUnits": 0,
              "secShowAxisTitle": true,
              "secAxisStyle": "showUnitOnly",
              "secColor": {"solid": {"color": "#949494"}},
              "secTitleText": "Add a Secondary Axis Title",
              "secTitleFontSize": 8,
              "secTitleFontFamily": "Arial",
              "secTitleBold": true,
              "secTitleItalic": false,
              "secTitleUnderline": false
          }],
      "legend": [{
              "show": true,
              "position": "Top",
              "showTitle": true,
              "titleText": "Add a legend title",
              "legendColor": {"solid": {"color": "#949494"}},
              "fontFamily": "Arial",
              "fontSize": 8,
              "bold": true,
              "italic": false,
              "underline": false,
              "legendMarkerRendering":"lineOnly",
              "matchLineColor":true,
              "defaultToCircle":true
              }],
    "zoom: [{... code not shown ...}],
      "smallMultiplesLayout: [{... code not shown ...}],
    "subHeader: [{... code not shown ...}],
  "labels": [{
          "show": true,
          "color": {"solid": {"color": "#ffffff"}},
          "labelDisplayUnits": 1,
          "labelPrecision": 2,
          "labelPosition": "InsideEnd",
          "labelOverflow": true,
          "fontFamily": "Arial",
          "fontSize": 7,
          "enableBackground": false,
          "backgroundColor": {"solid": {"color": "#949494"}},
          "backgroundTransparency": 0
          }],
    "lineStyles": [{
          "strokeWidth": 1,
          "strokeLineJoin": "Bevel",
```

```
        "lineStyle": "Solid",
        "showMarker": true,
        "markerShape": "longDash",
        "markerSize": 3,
        "markerColor":  {"solid": {"color": "#949494"}},
        "stepped": true,
        "showSeries": false
      }],
  "plotArea": [{... code not shown ...}],
  "title": [{... code not shown ...}],
  "subTitle": [{... code not shown ...}],
  "divider": [{... code not shown ...}],
  "spacing": [{... code not shown ...}],
  "background": [{... code not shown ...}],
  "lockAspect": [{... code not shown ...}],
  "border": [{... code not shown ...}],
  "dropShadow": [{... code not shown ...}],
  "visualHeader": [{... code not shown ...}]
  "visualHeaderTooltip": [{... code not shown ...}]
  "padding": [{... code not shown ...}]
  "visualTooltip: [{... code not shown ...}]
  }
}
```

Note that for the secondary axis, the only values that are accepted are 0. This means that the keyword "auto" cannot be entered as is the case for the main Y axis. Hopefully, this will be corrected by the time that this book is in print.

Tooltip options are described in Table 6-5.

Outline options are described in Table 6-10.

Legend keywords are given in Table 7-2.

Categories keywords are given in Table 7-3.

Values keywords are given in Table 7-4.

Zoom keywords are given in Table 7-5. For a column chart, the showOnValueAxis refers to the Y axis.

Labels keywords are given in Table 7-6.

Totals keywords are given in Table 7-7.

Plot area keywords are given in Table 7-8.

Small multiples options are described in Table 7-9.

Small multiples titles options are described in Table 7-10.

Legend position options are described in Table 7-11.

Category axis position options are described in Table 7-12.

Value axis display units options are described in Table 7-13.

Category axis display units options are described in Table 7-13.

Label position options are described in Table 7-14.

Gridline options are described in Table 7-15.

Value axis gridline style options are described in Table 7-16.

Line style options are described in Table 7-29.

The full JSON code to describe the formatting attributes of an area chart is available in the sample code files as Attributes_AreaChart.json.

Conclusion

In this chapter, you have seen how to prepare JSON themes to apply standard formatting to six core chart types – stacked bar chart, stacked column chart, clustered bar chart, clustered column chart, line chart, and area chart.

You have seen that they share many common features, which leads to a largely shared base of JSON code for format definitions.

The next step is to look in depth at half a dozen chart formatting specifications that are a little more complex – but that build on the JSON definitions that you have seen in this chapter.

CHAPTER 8

■ ■ ■

Complex Chart Visual Styles

The six chart types that you will be looking at in this chapter are only slightly more complex than the charts that you saw in the previous chapter. However, they are often more detailed (in most cases, anyway) than the charts that you saw previously. Fortunately, any added complexity is compensated by the fact that the JSON that you will see here is always built on – and derived from – the JSON that you discovered in the previous chapter.

The chart types whose formatting can be preset using JSON themes that you will look at in this chapter are

- 100% stacked bar chart

- 100% stacked column chart

- Stacked area chart

- Line and stacked column chart

- Line and clustered column chart

- Scatter chart

As ever, each chart type has a separate JSON file in the samples folder that contains the code that you can use in your own theme files.

100% Stacked Bar Chart

As you move on to looking at variations on a theme of bar and column charts, you will see that there are more similarities than differences in the JSON. In fact, the formatting for a 100% stacked bar chart is almost completely identical to the code that you use for a stacked bar chart.

The only differences compared to the JSON for a stacked bar chart are

- The keyword that specifies the formatting attributes of a 100% stacked bar chart. In this case, it is *hundredPercentStackedBarChart*.

- The absence of a *totals* section.

- The value axis no longer has an *axisScale* element.

Table 8-1 outlines the JSON keywords that you can use in stacked bar chart formatting and the section that corresponds to each keyword in the Power BI Desktop format pane.

© Adam Aspin 2023
A. Aspin, *Pro Power BI Theme Creation*, https://doi.org/10.1007/978-1-4842-9633-2_8

Table 8-1. *Stacked Bar Chart Keywords*

Keyword	Formatting Section
legend	Legend
categoryAxis	X Axis
valueAxis	Y Axis
zoom	Zoom slider
smallMultiplesLayout	Small Multiples
subheader	Small Multiples/Title
labels	Data labels
plotArea	Plot area
title	Title/Title
subTitle	Title/Subtitle
divider	Title/Divider
spacing	Title/Spacing
background	Effects/Background
border	Effects/Visual border
dropShadow	Effects/Shadow
visualHeader	Header Icons
lockAspect	Lock aspect
visualHeaderTooltip	Header Icons/Help tooltip
visualTooltip	Tooltips

The JSON code for a 100% stacked bar chart is available as the sample file Attributes_ HundredPercentStackedBarChart.json.

The following code snippet provides the JSON code used to pre-format a 100% stacked bar chart:

```
"hundredPercentStackedBarChart":
  {"*":
    {
    "general": [{... code not shown ...}]
    "legend": [{... code not shown ...}],
      "categoryAxis": [{
            "show": true,
            "position": "Left",
            "color": {"solid": {"color": "#949494"}},
            "fontSize": 8,
            "fontFamily": "Arial",
            "fontFamily": "Arial",
            "bold": true,
            "italic": false,
```

```
      "invertAxis": false,
      "concatenateLabels": true,
      "preferredCategoryWidth": 20,
      "maxMarginFactor": 25,
      "innerPadding": 15,
      "showAxisTitle": true,
      "axisStyle": "showTitleOnly",
      "titleColor": {"solid": {"color": "#949494"}},
      "titleText": "Add a Category Axis Title",
      "titleFontSize": 8,
      "titleFontFamily": "Arial",
      "titleBold": true,
      "titleItalic": false,
      "titleUnderline": false,
      "concatenateLabels": true,
      "switchAxisPosition": false
      }],
"valueAxis": [{
      "show": true,
      "start": "Auto",
      "end": "Auto",
      "labelColor": {"solid": {"color": "#949494"}},
      "fontSize": 8,
      "fontFamily": "Arial",
      "fontFamily": "Arial",
      "bold": true,
      "italic": false,
      "labelPrecision": 0,
      "labelDisplayUnits": 0,
      "showAxisTitle": true,
      "axisStyle": "showTitleOnly",
      "color": {"solid": {"color": "#949494"}},
      "titleText": "Add a Value Axis Title",
      "titleFontSize": 8,
      "titleFontFamily": "Arial",
      "titleBold": true,
      "titleItalic": false,
      "titleUnderline": false,
      "gridLineShow": false,
      "gridLineColor":  {"solid": {"color": "#949494"}},
      "gridLineThickness": 1,
      "gridlineStyle": "solid",
      "invertAxis": true
    }],
"labels": [{
      "show": true,
      "color": {"solid": {"color": "#ffffff"}},
      "labelDisplayUnits": 1,
      "labelPrecision": 2,
      "labelPosition": "InsideEnd",
      "labelOverflow": true,
```

```
            "fontFamily": "Arial",
            "fontSize": 7,
            "enableBackground": false,
            "backgroundColor": {"solid": {"color": "#949494"}},
            "backgroundTransparency": 0
            }],
    "zoom[{... code not shown ...}],
    "smallMultiplesLayout: [{... code not shown ...}],
    "subHeader: [{... code not shown ...}],
    "plotArea": [{... code not shown ...}],
    "title": [{... code not shown ...}],
    "subTitle": [{... code not shown ...}],
    "divider": [{... code not shown ...}],
    "spacing": [{... code not shown ...}],
    "background": [{... code not shown ...}],
    "lockAspect": [{... code not shown ...}],
    "border": [{... code not shown ...}],
    "dropShadow": [{... code not shown ...}],
    "visualHeader": [{... code not shown ...}]
    "visualHeaderTooltip": [{... code not shown ...}]
    "padding": [{... code not shown ...}]
    "visualTooltip: [{... code not shown ...}]
    }
}
```

Tooltip options are described in Table 6-5.

Legend keywords are given in Table 7-2.

Categories keywords are given in Table 7-3.

Values keywords are given in Table 7-4.

Zoom keywords are given in Table 7-5. For a column chart, the showOnValueAxis refers to the Y axis.

Labels keywords are given in Table 7-6.

Totals keywords are given in Table 7-7.

Plot area keywords are given in Table 7-8.

Small multiples options are described in Table 7-9.

Small multiples titles options are described in Table 7-10.

Legend position options are described in Table 7-11.

Category axis position options are described in Table 7-12.

Value axis display units options are described in Table 7-13.

Category axis display units options are described in Table 7-13.

Label position options are described in Table 7-14.

Gridline options are described in Table 7-15.

Value axis gridline style options are described in Table 7-16.

100% Stacked Column Chart

The formatting for a 100% stacked column chart is nearly identical to the code that you use for a stacked column chart. The essential differences are

- The keyword that specifies the formatting attributes of a 100% stacked column chart is *hundredPercentStackedColumnChart*.

- There is no *totals* section.

- There is no *axisScale* in the values axis.

Table 8-2 outlines the JSON keywords that you can use in stacked column chart formatting and the section that corresponds to each keyword in the Power BI Desktop format pane.

Table 8-2. *Stacked Column Chart Keywords*

Keyword	Formatting Section
legend	Legend
categoryAxis	X Axis
valueAxis	Y Axis
zoom	Zoom slider
smallMultiplesLayout	Small Multiples
subheader	Small Multiples/Title
labels	Data labels
plotArea	Plot area
title	Title/Title
subTitle	Title/Subtitle
divider	Title/Divider
spacing	Title/Spacing
background	Effects/Background
border	Effects/Visual border
dropShadow	Effects/Shadow
visualHeader	Header Icons/Icons Header Icons/Colors
lockAspect	Lock aspect
visualHeaderTooltip	Header Icons/Help tooltip
visualTooltip	Tooltips

The JSON code for a 100% stacked column chart is available in the sample file Attributes_HundredPercentStackedColumnChart.json.

The following code snippet provides the JSON code used to pre-format a 100% stacked column chart:

```
"hundredPercentStackedColumnChart":
  {"*":
    {
    "legend": [{... code not shown ...}],
      "categoryAxis": [{
```

```
          "show": true,
          "color": {"solid": {"color": "#949494"}},
          "fontSize": 8,
          "fontFamily": "Arial",
          "bold": true,
          "italic": false,
          "underline": false,
          "preferredCategoryWidth": 20,
          "maxMarginFactor": 25,
          "innerPadding": 15,
          "concatenateLabels": true,
          "showAxisTitle": true,
          "axisStyle": "showTitleOnly",
          "titleColor": {"solid": {"color": "#949494"}},
          "titleText": "Add an Axis Title",
          "titleFontSize": 8,
          "titleFontFamily": "Arial",
          "titleBold": true,
          "titleItalic": false,
          "titleUnderline": false,
          "concatenateLabels": true
          }],
     "valueAxis": [{
          "show": true,
          "position": "Left",
          "start": 0,
          "end": 0,
          "labelColor": {"solid": {"color": "#949494"}},
          "fontSize": 8,
          "fontFamily": "Arial",
          "bold": true,
          "italic": false,
          "underline": false,
          "labelPrecision": 0,
          "labelDisplayUnits": 0,
          "showAxisTitle": true,
          "axisStyle": "showTitleOnly",
          "color": {"solid": {"color": "#949494"}},
          "titleText": "Add an Axis Title",
          "titleFontSize": 8,
          "titleFontFamily": "Arial",
          "titleBold": true,
          "titleItalic": false,
          "titleUnderline": false,
          "gridLineShow": false,
          "gridLineColor": {"solid": {"color": "#949494"}},
          "gridLineThickness": 1,
          "gridlineStyle": "solid",
          "invertAxis": true
          }],
     "zoom[{... code not shown ...}],
```

```
"smallMultiplesLayout: [{... code not shown ...}],
"subHeader: [{... code not shown ...}],
 "labels": [{
        "show": true,
        "color": {"solid": {"color": "#ffffff"}},
        "labelDisplayUnits": 1,
        "labelPrecision": 2,
        "labelOrientation": 0,
        "labelPosition": "InsideEnd",
        "labelOverflow": true,
        "fontFamily": "Arial",
        "fontSize": 7,
        "enableBackground": false,
        "backgroundColor": {"solid": {"color": "#949494"}},
        "backgroundTransparency": 0
        }],
"zoom[{... code not shown ...}],
"smallMultiplesLayout: [{... code not shown ...}],
"subHeader: [{... code not shown ...}],
"plotArea": [{... code not shown ...}],
"title": [{... code not shown ...}],
"subTitle": [{... code not shown ...}],
"divider": [{... code not shown ...}],
"spacing": [{... code not shown ...}],
"background": [{... code not shown ...}],
"lockAspect": [{... code not shown ...}],
"border": [{... code not shown ...}],
"dropShadow": [{... code not shown ...}],
"visualHeader": [{... code not shown ...}]
"visualHeaderTooltip": [{... code not shown ...}]
"padding": [{... code not shown ...}]
"visualTooltip: [{... code not shown ...}]
}
}
```

Tooltip options are described in Table 6-5.

Legend keywords are given in Table 7-2.

Categories keywords are given in Table 7-3.

Values keywords are given in Table 7-4.

Zoom keywords are given in Table 7-5. For a column chart, the showOnValueAxis refers to the Y axis.

Labels keywords are given in Table 7-6.

Totals keywords are given in Table 7-7.

Plot area keywords are given in Table 7-8.

Small multiples options are described in Table 7-9.

Small multiples titles options are described in Table 7-10.

Legend position options are described in Table 7-11.

Category axis position options are described in Table 7-12.

Value axis display units options are described in Table 7-13.

Category axis display units options are described in Table 7-13.

Label position options are described in Table 7-14.

Gridline options are described in Table 7-15.

Value axis gridline style options are described in Table 7-16.

Stacked Area Chart

A stacked area chart is, essentially, an extension of the area chart – but without the secondary Y axis.

The keyword that specifies the formatting attributes of a stacked area chart is *stackedAreaChart*.

Table 8-3 outlines the JSON keywords that you can use in stacked area chart formatting and the section that corresponds to each keyword in the Power BI Desktop format pane.

Table 8-3. *Stacked Area Chart Keywords*

Keyword	Formatting Section
legend	Legend
categoryAxis	X Axis
valueAxis	Y Axis
zoom	Zoom slider
smallMultiplesLayout	Small Multiples
subheader	Small Multiples/Title
labels	Data labels
seriesLabels	Series labels
zoom	Zoom Slider
plotArea	Plot area
title	Title/Title
subTitle	Title/Subtitle
divider	Title/Divider
spacing	Title/Spacing
background	Effects/Background
border	Effects/Visual border
dropShadow	Effects/Shadow
visualHeader	Header Icons/Icons Header Icons/Colors
lockAspect	Lock aspect
visualHeaderTooltip	Header Icons/Help tooltip
visualTooltip	Tooltips

The full JSON code to describe the formatting attributes of a stacked area chart is available in Attributes_StackedAreaChart.json in the sample files.

The following is the JSON that describes the formatting for a stacked area chart:

```
"stackedAreaChart":
{"*" :
  {
    "categoryAxis": [{
        "show": true,
        "axisType":"Categorical",
        "color": {"solid": {"color": "#949494"}},
        "fontSize": 8,
        "fontFamily": "Arial",
        "bold": true,
        "italic": false,
        "underline": false,
        "preferredCategoryWidth": 20,
        "maxMarginFactor": 30,
        "concatenateLabels": true,
        "showAxisTitle": true,
        "axisStyle": "showTitleOnly",
        "titleColor": {"solid": {"color": "#949494"}},
        "titleText": "Add an Axis Title",
        "titleFontSize": 8,
        "titleFontFamily": "Arial",
        "concatenateLabels": true
        }],
    "valueAxis": [{
        "show": true,
        "position": "Left",
        "axisScale": "linear",
        "start": "Auto",
        "end": "Auto",
        "labelColor": {"solid": {"color": "#949494"}},
        "fontSize": 8,
        "fontFamily": "Arial",
        "bold": true,
        "italic": false,
        "underline": false,
        "labelPrecision": 0,
        "labelDisplayUnits": 0,
        "showAxisTitle": true,
        "axisStyle": "showUnitOnly",
        "color": {"solid": {"color": "#949494"}},
        "titleText": "Add an Axis Title",
        "titleFontSize": 8,
        "titleFontFamily": "Arial",
        "gridLineShow": true,
        "gridLineColor":  {"solid": {"color": "#949494"}},
        "gridLineThickness": 2,
```

155

```
                "gridlineStyle": "dashed",
                "invertAxis": true
            }],
    "legend": [{
                "show": true,
                "position": "Top",
                "showTitle": true,
                "titleText": "Add a legend title",
                "legendColor": {"solid": {"color": "#949494"}},
                "fontFamily": "Arial",
                "fontSize": 8,
                "bold": true,
                "italic": false,
                "underline": false,
                "legendMarkerRendering":"lineOnly"
            }],
    "labels": [{
                "show": true,
                "color": {"solid": {"color": "#ffffff"}},
                "labelDisplayUnits": 1,
                "labelPrecision": 2,
                "labelOrientation": 1,
                "labelPosition": "InsideEnd",
                "labelOverflow": true,
                "fontFamily": "Arial",
                "fontSize": 7,
                "bold": true,
                "italic": false,
                "underline": false,
                "enableBackground": false,
                "backgroundColor": {"solid": {"color": "#949494"}},
                "backgroundTransparency": 0
            }],
    "seriesLabels": [{
                "show": true,
                "seriesMaximumWidth": 1,
                "seriesPosition": "Left",
                "seriesFontFamily": "Arial",
                "textSize": 9,
                "bold": false,
                "italic": false,
                "underline": false,
                "seriesColor": {"solid": {"color": "#949494"}},
                "seriesWordWrap": false,
                "enableBackground": false,
                "backgroundColor": {"solid": {"color": "#12239E"}},
                "backgroundTransparency": 0
            }],
    "totals": [{
                "show": true,
                "color": {"solid": {"color": "#FFFFFF"}},
```

```
            "labelDisplayUnits": 1000,
            "labelPrecision": 2,
            "fontSize": 8,
            "fontFamily": "Arial",
            "bold": true,
            "italic": false,
            "underline": false,
            "enableBackground": true,
            "backgroundColor": {"solid": {"color": "#949494"}},
            "backgroundTransparency": 10,
            "showPositiveAndNegative": true
            }],
      "lineStyles": [{
            "strokeWidth": 1,
            "strokeLineJoin": "bevel",
            "lineStyle": "dashed",
            "showMarker": true,
            "markerShape": "longDash",
            "markerSize": 3,
            "markerColor":  {"solid": {"color": "#949494"}},
            "stepped": true,
            "showSeries": false
        }],
    "zoom[{... code not shown ...}],
    "smallMultiplesLayout: [{... code not shown ...}],
    "subHeader: [{... code not shown ...}],
    "plotArea": [{... code not shown ...}],
    "title": [{... code not shown ...}],
    "subTitle": [{... code not shown ...}],
    "divider": [{... code not shown ...}],
    "spacing": [{... code not shown ...}],
    "background": [{... code not shown ...}],
    "lockAspect": [{... code not shown ...}],
    "border": [{... code not shown ...}],
    "dropShadow": [{... code not shown ...}],
    "visualHeader": [{... code not shown ...}]
    "visualHeaderTooltip": [{... code not shown ...}]
    "padding": [{... code not shown ...}]
    "visualTooltip: [{... code not shown ...}]
    }
}
```

The *seriesLabels* keyword that describes the Series labels section uses the keywords shown in Table 8-4.

Table 8-4. *Keyword Mapping for Series Labels*

Element	Keyword
Series labels	Show
Options/Maximum width	seriesMaximumWidth
Options/Series position	seriesPosition
Values/Font	seriesFontFamily
Values/(Size)	textSize
Values/(bold)	Bold
Values/(italic)	Italic
Values/(underline)	Underline
Values/Series color	seriesColor
Values/Wordwrap	seriesWordWrap
Background	enableBackground
Background/Background color	backgroundColor
Background/Transparency	backgroundTransparency

Tooltip options are described in Table 6-5.
Legend keywords are given in Table 7-2.
Categories keywords are given in Table 7-3.
Values keywords are given in Table 7-4.
Zoom keywords are given in Table 7-5. For a column chart, the showOnValueAxis refers to the Y axis.
Labels keywords are given in Table 7-6.
Totals keywords are given in Table 7-7.
Plot area keywords are given in Table 7-8.
Small multiples options are described in Table 7-9.
Small multiples titles options are described in Table 7-10.
Legend position options are described in Table 7-11.
Category axis position options are described in Table 7-12.
Value axis display units options are described in Table 7-13.
Category axis display units options are described in Table 7-13.
Label position options are described in Table 7-14.
Gridline options are described in Table 7-15.
Value axis gridline style options are described in Table 7-16.
Series labels options are described in Table 7-27.
Series label position options are described in Table 7-28.
Line style options are described in Table 7-29.
Join type options are described in Table 7-33.

Line and Stacked Column Chart

One of the two combined chart type options is the line and stacked column chart. As far as the JSON theme code is concerned, it is a mixture of the column chart and the line chart.

The keyword that specifies the formatting attributes of a line and stacked column chart is *lineStackedColumnComboChart*.

The essential differences as far as the JSON is concerned are in the value axis – where the multiple series are differentiated.

Table 8-5 outlines the JSON keywords that you can use in line and stacked column chart formatting and the section that corresponds to each keyword in the Power BI Desktop format pane.

Table 8-5. *Line and Stacked Column Chart Keywords*

Keyword	Formatting Section
legend	Legend
categoryAxis	X Axis
valueAxis	Y Axis
zoom	Zoom slider
labels	Data labels
seriesLabels	Series Labels
lineStyles	Shapes
smallMultiplesLayout	Small Multiples
subheader	Small Multiples/Title
zoom	Zoom Slider
plotArea	Plot area
title	Title/Title
subTitle	Title/Subtitle
divider	Title/Divider
spacing	Title/Spacing
background	Effects/Background
border	Effects/Visual border
dropShadow	Effects/Shadow
visualHeader	Header Icons/Icons Header Icons/Colors
lockAspect	Lock aspect
visualHeaderTooltip	Header Icons/Help tooltip
visualTooltip	Tooltips

The following is the JSON that describes the formatting for a line and stacked column chart. It is one of the longest JSON descriptions for any visual.

```
"lineStackedColumnComboChart":
  {"*":
    {
      "categoryAxis": [{
          "show": true,
          "color": {"solid": {"color": "#949494"}},
          "fontSize": 8,
          "fontFamily": "Arial",
          "bold": true,
          "italic": false,
          "underline": false,
          "preferredCategoryWidth": 20,
          "maxMarginFactor": 25,
          "innerPadding": 15,
          "concatenateLabels": true,
          "showAxisTitle": true,
          "axisStyle": "showTitleOnly",
          "titleColor": {"solid": {"color": "#949494"}},
          "titleText": "Add an Axis Title",
          "titleFontSize": 8,
          "titleFontFamily": "Arial",
          "titleBold": true,
          "titleItalic": false,
          "titleUnderline": false,
          "concatenateLabels": true
          }],
      "valueAxis": [{
          "show": true,
          "position": "Left",
          "start": "Auto",
          "end": "Auto",
          "labelColor": {"solid": {"color": "#949494"}},
          "fontSize": 8,
          "fontFamily": "Arial",
          "bold": true,
          "italic": false,
          "underline": false,
          "labelPrecision": 0,
          "labelDisplayUnits": 0,
          "showAxisTitle": true,
          "axisStyle": "showTitleOnly",
          "color": {"solid": {"color": "#949494"}},
          "titleText": "Add an Axis Title",
          "titleFontSize": 8,
          "titleFontFamily": "Arial",
          "titleBold": true,
          "titleItalic": false,
          "titleUnderline": false,
          "gridLineShow": false,
          "gridLineColor": {"solid": {"color": "#949494"}},
          "gridLineThickness": 1,
          "gridlineStyle": "solid",
```

```
    "invertAxis": true,
    "secShow": true,
    "alignZeros": true,
    "secPosition": "Right",
    "secStart": 0,
    "secEnd": 0,
    "secAxisScale": "linear",
    "secLabelColor": {"solid": {"color": "#949494"}},
    "secFontSize": 8,
    "secFontFamily": "Arial",
    "secBold": true,
    "secItalic": false,
    "secUnderline": false,
    "secLabelDisplayUnits": 0,
    "secLabelPrecision": 0,
    "secShowAxisTitle": true,
    "secTitleText": "Add an Axis Title",
    "secAxisStyle": "showTitleOnly",
    "secTitleColor": {"solid": {"color": "#949494"}},
    "secTitleText": "Add an Axis Title",
    "secTitleFontSize": 8,
    "secTitleFontFamily": "Arial",
    "secTitleBold": true,
    "secTitleItalic": false,
    "secTitleUnderline": false
    }],
"labels": [{
    "show": true,
    "color": {"solid": {"color": "#ffffff"}},
    "labelDisplayUnits": 1,
    "labelPrecision": 2,
    "labelOrientation": 1,
    "labelPosition": "InsideEnd",
    "labelOverflow": true,
    "fontFamily": "Arial",
    "fontSize": 7,
    "bold": true,
    "italic": false,
    "underline": false,
    "enableBackground": false,
    "backgroundColor": {"solid": {"color": "#949494"}},
    "backgroundTransparency": 0
    }],
"seriesLabels": [{
    "show": true,
    "seriesMaximumWidth": 1,
    "seriesPosition": "Left",
    "seriesFontFamily": "Arial",
    "textSize": 9,
    "bold": false,
    "italic": true,
    "underline": false,
```

```
            "seriesColor": {"solid": {"color": "#949494"}},
            "seriesWordWrap": false,
            "enableBackground": false,
            "backgroundColor": {"solid": {"color": "#12239E"}},
            "backgroundTransparency": 0
            }],
      "lineStyles": [{
            "strokeWidth": 1,
            "strokeLineJoin": "bevel",
            "lineStyle": "solid",
            "showMarker": true,
            "markerShape": "longDash",
            "markerSize": 3,
            "markerColor":  {"solid": {"color": "#949494"}},
            "stepped": true,
            "showSeries": false
         }],
      "totals": [{
            "show": true,
            "color": {"solid": {"color": "#FFFFFF"}},
            "labelDisplayUnits": 1000,
            "labelPrecision": 2,
            "fontSize": 8,
            "fontFamily": "Arial",
            "bold": true,
            "italic": false,
            "underline": false,
            "enableBackground": true,
            "backgroundColor": {"solid": {"color": "#949494"}},
            "backgroundTransparency": 10,
            "showPositiveAndNegative": true
            }],
   "seriesLabels[{... code not shown ...}],
   "totals[{... code not shown ...}],
   "zoom[{... code not shown ...}],
   "smallMultiplesLayout: [{... code not shown ...}],
   "subHeader: [{... code not shown ...}],
   "plotArea": [{... code not shown ...}],
   "title": [{... code not shown ...}],
   "subTitle": [{... code not shown ...}],
   "divider": [{... code not shown ...}],
   "spacing": [{... code not shown ...}],
   "background": [{... code not shown ...}],
   "lockAspect": [{... code not shown ...}],
   "border": [{... code not shown ...}],
   "dropShadow": [{... code not shown ...}],
   "visualHeader": [{... code not shown ...}]
   "visualHeaderTooltip": [{... code not shown ...}]
   "padding": [{... code not shown ...}]
   "visualTooltip": [{... code not shown ...}]
   }
 }
```

Legend keywords are given in Table 7-2. It contains an additional option – *legendMarkerRendering*.

Categories keywords are given in Table 7-3. They also contain the *concatenateLabels* and *showAxisTitle* elements.

The *valueAxis* keyword that describes the Y axis section uses the keywords shown in Table 8-6.

Table 8-6. *Keyword Mapping for the Y Axis*

Element	Keyword
Y axis (Y axis column)	Show
(Y axis column)	Position
Scale type (Y axis column)	axisScale
Start (Y axis column)	Start
End (Y axis column)	End
Color (Y axis column)	labelColor
Text size (Y axis column)	fontSize
Font family (Y axis column)	fontFamily
(Bold)	Bold
(Italic)	Italic
(Underline)	Underline
Display units (Y axis column)	labelDisplayUnits
Value decimal places (Y axis column)	labelPrecision
Title (Y axis column)	showAxisTitle
Style (Y axis column)	axisStyle
Title color (Y axis column)	Color
Axis title (Y axis column)	titleText
Title text size (Y axis column)	titleFontSize
(Title) Font family (Y axis column)	titleFontFamily
Title/(Bold)	titleBold
Title/(Italic)	titleItalic
Title/(Underline)	titleUnderline
Gridlines (Y axis column)	gridLineShow
(Gridline) Color (Y axis column)	gridLineColor
Stroke width (Y axis column)	gridLineThickness
Line style (Y axis column)	gridlineStyle

(continued)

Table 8-6. (*continued*)

Element	Keyword
Show secondary	secShow
Align zeros	alignZeros
(Y axis row)	secPosition
Scale type (Y axis row)	secAxisScale
Start (Y axis row)	secStart
End (Y axis row)	secEnd
Color (Y axis row)	secLabelColor
Text size (Y axis row)	secFontSize
Font family (Y axis row)	secFontFamily
(Bold)	secBold
(Italic)	secItalic
(Underline)	secUnderline
Display units (Y axis row)	secLabelDisplayUnits
Value decimal places (Y axis row)	secLabelPrecision
Title (Y axis row)	secShowAxisTitle
Style (Y axis row)	secAxisStyle
Title color (Y axis row)	secTitleColor
Axis title (Y axis row)	secTitleText
Title text size (Y axis row)	secTitleFontSize
(Title) Font family (Y axis row)	secTitleFontFamily
Title/(Bold) (Y axis row)	secTitleBold
Title/(Italic) (Y axis row)	secTitleItalic
Title/(Underline) (Y axis row)	secTitleUnderline
Invert range	invertAxis

Zoom keywords are given in Table 7-5.

Labels keywords are given in Table 7-6. This also contains the *labelOrientation* option.

Line style options are described in Table 7-29.

Tooltip keywords are given in Table 6-5.

The *legendMarkerRendering* keyword in the JSON in the legend section for a line and stacked column chart needs one of the attributes shown in Table 8-7.

Table 8-7. *Legend Position Options*

Value	Formatting Popup Option
markerOnly	Marker only
lineAndMarker	Line and markers
lineOnly	Line only

Tooltip options are described in Table 6-5.
Legend keywords are given in Table 7-2.
Categories keywords are given in Table 7-3.
Values keywords are given in Table 7-4.
Zoom keywords are given in Table 7-5. For a column chart, the showOnValueAxis refers to the Y axis.
Labels keywords are given in Table 7-6.
Totals keywords are given in Table 7-7.
Plot area keywords are given in Table 7-8.
Small multiples options are described in Table 7-9.
Small multiples titles options are described in Table 7-10.
Legend position options are described in Table 7-11.
Category axis position options are described in Table 7-12.
Value axis display units options are described in Table 7-13.
Category axis display units options are described in Table 7-13.
Label position options are described in Table 7-14.
Gridline options are described in Table 7-15.
Value axis gridline style options are described in Table 7-16.
Series labels options are described in Table 7-27.
Series label position options are described in Table 7-28.
Line style options are described in Table 7-29.
Join type options are described in Table 7-33.
The full JSON code to describe the formatting attributes of a line and stacked column chart is in the file Attributes_LineAndStackedColumnChart.json.

Line and Clustered Column Chart

The other combined chart type option is the line and clustered column chart. The JSON code for this visual is very similar to the code for the line and stacked column chart.

The keyword that specifies the formatting attributes of a line and clustered column chart is *lineClusteredColumnComboChart*.

Table 8-8 outlines the JSON keywords that you can use in line and clustered column chart formatting and the section that corresponds to each keyword in the Power BI Desktop format pane.

Table 8-8. *Line and Clustered Column Chart Keywords*

Keyword	Formatting Section
legend	Legend
categoryAxis	X Axis
valueAxis	Y Axis
categoryLabels	Category labels
plotArea	Plot area
seriesLabels	Series Labels
lineStyles	Lines
smallMultiplesLayout	Small Multiples
subheader	Small Multiples/Title
zoom	Zoom slider
title	Title/Title
subTitle	Title/Subtitle
divider	Title/Divider
spacing	Title/Spacing
background	Effects/Background
border	Effects/Visual border
dropShadow	Effects/Shadow
visualHeader	Header Icons
lockAspect	Lock aspect
visualHeaderTooltip	Header Icons/Help tooltip
visualTooltip	Tooltips

The following is the JSON that describes the formatting for a line and clustered column chart:

```
"lineClusteredColumnComboChart":
  {"*":
    {
    "general": [{... code not shown ...}]
      "categoryAxis": [{
            "show": true,
            "color": {"solid": {"color": "#949494"}},
            "fontSize": 8,
            "fontFamily": "Arial",
            "bold": true,
            "italic": false,
            "underline": false,
```

```
        "preferredCategoryWidth": 20,
        "maxMarginFactor": 25,
        "innerPadding": 15,
        "concatenateLabels": true,
        "showAxisTitle": true,
        "axisStyle": "showTitleOnly",
        "titleColor": {"solid": {"color": "#949494"}},
        "titleText": "Add an Axis Title",
        "titleFontSize": 8,
        "titleFontFamily": "Arial",
        "concatenateLabels": true
        }],
"valueAxis": [{
        "show": true,
        "position": "Left",
        "start": "Auto",
        "end": "Auto",
        "labelColor": {"solid": {"color": "#949494"}},
        "fontSize": 8,
        "fontFamily": "Arial",
        "bold": true,
        "italic": false,
        "underline": false,
        "labelPrecision": 0,
        "labelDisplayUnits": 0,
        "showAxisTitle": true,
        "axisStyle": "showTitleOnly",
        "color": {"solid": {"color": "#949494"}},
        "titleText": "Add an Axis Title",
        "titleFontSize": 8,
        "titleFontFamily": "Arial",
        "titleBold": true,
        "titleItalic": false,
        "titleUnderline": false,
        "gridLineShow": false,
        "gridLineColor":  {"solid": {"color": "#949494"}},
        "gridLineThickness": 1,
        "gridlineStyle": "solid",
        "invertAxis": true,
        "secShow": true,
        "alignZeros": true,
        "secPosition": "Right",
        "secStart": 0,
        "secEnd": 0,
        "secAxisScale": "linear",
        "secLabelColor": {"solid": {"color": "#949494"}},
        "secFontSize": 8,
        "secFontFamily": "Arial",
        "secBold": true,
        "secItalic": false,
        "secUnderline": false,
```

```
            "secLabelDisplayUnits": 0,
            "secLabelPrecision": 0,
            "secShowAxisTitle": true,
            "secTitleText": "Add a secondary Axis Title",
            "secAxisStyle": "showTitleOnly",
            "secTitleColor": {"solid": {"color": "#949494"}},
            "secTitleFontSize": 8,
            "secTitleFontFamily": "Arial",
            "secTitleBold": true,
            "secTitleItalic": false,
            "secTitleUnderline": false
        }],
    "legend": [{
            "show": true,
            "position": "Top",
            "showTitle": true,
            "titleText": "Add a legend title",
            "legendColor": {"solid": {"color": "#949494"}},
            "fontFamily": "Arial",
            "fontSize": 8,
            "bold": true,
            "italic": false,
            "underline": false,
            "legendMarkerRendering":"markerOnly"
        }],
    "lineStyles": [{
            "shadeArea": true,
            "strokeWidth": 1,
            "strokeLineJoin": "bevel",
            "lineStyle": "solid",
            "showMarker": true,
            "stepped": true
        }],
    "seriesLabels": [{
            "show": true,
            "seriesMaximumWidth": 1,
            "seriesPosition": "Left",
            "seriesFontFamily": "Arial",
            "textSize": 9,
            "bold": false,
            "italic": true,
            "underline": false,
            "seriesColor": {"solid": {"color": "#949494"}},
            "seriesWordWrap": false,
            "enableBackground": false,
            "backgroundColor": {"solid": {"color": "#12239E"}},
            "backgroundTransparency": 0
        }],
    "zoom[{... code not shown ...}],
    "smallMultiplesLayout: [{... code not shown ...}],
    "subHeader: [{... code not shown ...}],
```

```
    "plotArea": [{... code not shown ...}],
    "title": [{... code not shown ...}],
    "subTitle": [{... code not shown ...}],
    "divider": [{... code not shown ...}],
    "spacing": [{... code not shown ...}],
    "background": [{... code not shown ...}],
    "lockAspect": [{... code not shown ...}],
    "border": [{... code not shown ...}],
    "dropShadow": [{... code not shown ...}],
    "visualHeader": [{... code not shown ...}]
    "visualHeaderTooltip": [{... code not shown ...}]
    "padding": [{... code not shown ...}]
    "visualTooltip: [{... code not shown ...}]
    }
}
```

You can find the JSON code for this in the sample file Attributes_LineAndClusteredColumnChart.json.

Tooltip options are described in Table 6-5.

Legend keywords are given in Table 7-2.

Categories keywords are given in Table 7-3.

Values keywords are given in Table 7-4.

Zoom keywords are given in Table 7-5. For a column chart, the showOnValueAxis refers to the Y axis.

Labels keywords are given in Table 7-6.

Totals keywords are given in Table 7-7.

Plot area keywords are given in Table 7-8.

Small multiples options are described in Table 7-9.

Small multiples titles options are described in Table 7-10.

Legend position options are described in Table 7-11.

Category axis position options are described in Table 7-12.

Value axis display units options are described in Table 7-13.

Category axis display units options are described in Table 7-13.

Label position options are described in Table 7-14.

Gridline options are described in Table 7-15.

Value axis gridline style options are described in Table 7-16.

Series labels options are described in Table 7-27.

Series label position options are described in Table 7-28.

Line style options are described in Table 7-29.

Join type options are described in Table 7-33.

Scatter Chart

The keyword that specifies the formatting attributes of a scatter chart is *scatterChart*.

Table 8-9 outlines the JSON keywords that you can use in scatter chart formatting and the section that corresponds to each keyword in the Power BI Desktop format pane.

Table 8-9. *Scatter Chart Keywords*

Keyword	Formatting Section
Legend	Legend
categoryAxis	X Axis
valueAxis	Y Axis
zoom	Zoom slider
categoryLabels	Category label
lineStyles	Shapes
plotArea	Plot area
bubbles	Markers
zoom	Zoom Slider
title	Title/Title
subTitle	Title/Subtitle
divider	Title/Divider
spacing	Title/Spacing
background	Effects/Background
border	Effects/Visual border
dropShadow	Effects/Shadow
visualHeader	Header Icons/Icons Header Icons/Colors
lockAspect	Lock aspect
visualHeaderTooltip	Header Icons/Help tooltip
visualTooltip	Tooltips

The following is the JSON that describes the formatting of a scatter chart:

```
"scatterChart":
  {"*":
    {
    "general": [{... code not shown ...}]
    "legend": [{... code not shown ...}],
      "categoryAxis": [{
            "show": true,
            "axisScale": "linear",
            "start": 0,
            "end": 0,
            "color": {"solid": {"color": "#949494"}},
            "fontSize": 8,
            "fontFamily": "Arial",
```

```
        "fontSize": 8,
        "fontFamily": "Arial",
        "bold": true,
        "italic": false,
        "underline": false,
        "preferredCategoryWidth": 20,
        "maxMarginFactor": 25,
        "showAxisTitle": true,
        "axisStyle": "showTitleOnly",
        "titleColor": {"solid": {"color": "#949494"}},
        "titleText": "Add an Axis Title",
        "titleFontSize": 8,
        "titleFontFamily": "Arial",
        "titleBold": true,
        "titleItalic": false,
        "titleUnderline": false,
        "gridLineShow": true,
        "gridLineColor":  {"solid": {"color": "#949494"}},
        "gridLineThickness": 1,
        "gridlineStyle": "solid",
        "treatNullsAsZero": true
        }],
"valueAxis": [{
        "show": true,
        "position": "Left",
        "start": 0,
        "end": 0,
        "labelColor": {"solid": {"color": "#949494"}},
        "fontSize": 8,
        "fontFamily": "Arial",
        "fontSize": 8,
        "fontFamily": "Arial",
        "bold": true,
        "italic": false,
        "underline": false,
        "labelPrecision": 0,
        "labelDisplayUnits": 0,
        "showAxisTitle": true,
        "axisStyle": "showTitleOnly",
        "color": {"solid": {"color": "#949494"}},
        "titleText": "Add an Axis Title",
        "titleFontSize": 8,
        "titleFontFamily": "Arial",
        "titleBold": true,
        "titleItalic": false,
        "titleUnderline": false,
        "invertAxis": true,
        "gridLineShow": false,
        "gridLineColor":  {"solid": {"color": "#949494"}},
        "gridLineThickness": 1,
        "gridlineStyle": "solid",
```

```
                "treatNullsAsZero": true
            }],
        "bubbles": [{
            "bubbleSize": 15,
            "markerShape": "square"
        }],
        "categoryLabels": [{
            "show": true,
            "fontColor": { "solid": { "color": "#000000"}},
            "fontSize": 11,
            "fontFamily": "Arial",
            "bold": true,
            "italic": false,
            "underline": false,
            "enableBackground": true,
            "backgroundColor": { "solid": { "color": "#949494"}},
            "backgroundTransparency": 25
        }],
    "zoom[{... code not shown ...}],
    "smallMultiplesLayout: [{... code not shown ...}],
    "subHeader: [{... code not shown ...}],
    "plotArea": [{... code not shown ...}],
    "title": [{... code not shown ...}],
    "subTitle": [{... code not shown ...}],
    "divider": [{... code not shown ...}],
    "spacing": [{... code not shown ...}],
    "lockAspect": [{... code not shown ...}],
    "border": [{... code not shown ...}],
    "dropShadow": [{... code not shown ...}],
    "visualHeader": [{... code not shown ...}]
    "visualHeaderTooltip": [{... code not shown ...}]
    "padding": [{... code not shown ...}]
    "visualTooltip: [{... code not shown ...}]
    }
}
```

Legend keywords are given in Table 7-2.

The *categoryAxis* keyword that describes the X axis section uses the keywords shown in Table 8-10.

Table 8-10. *Keyword Mapping for the X Axis*

Element	Keyword
X axis	Show
Scale type	axisScale
Start	Start
End	End
Color	labelColor
Text size	fontSize
Font family	fontFamily
Display units	labelDisplayUnits
Value decimal places	labelPrecision
Title	showAxisTitle
Gridlines	gridLineShow
(Gridline) Color	gridLineColor
Stroke width	gridLineThickness
Line style	gridlineStyle
Show blank values	treatNullsAsZero
Invert Range	invertAxis

The *valueAxis* keyword that describes the Y axis section uses the keywords shown in Table 8-11.

Table 8-11. *Keyword Mapping for the Y Axis*

Element	Keyword
Y axis	Show
Position	Position
Scale type	axisScale
Start	Start
End	End
Color	labelColor
Text size	fontSize
Font family	fontFamily
Display units	labelDisplayUnits
Value decimal places	labelPrecision
Title	showAxisTitle
Style	axisStyle
Title color	Color
Axis title	titleText
Title text size	titleFontSize
(Title) Font family	titleFontFamily
Gridlines	gridLineShow
(Gridline) Color	gridLineColor
Stroke width	gridLineThickness
Line style	gridlineStyle
Show blank values	treatNullsAsZero

Zoom keywords are given in Table 7-5.
Labels keywords are given in Table 7-6.
The *bubbles* keyword that describes the shapes section uses the keywords shown in Table 8-12.

Table 8-12. *Keyword Mapping for the Shapes*

Element	Keyword
Shapes	Show
Size	bubbleSize
Marker shape	markerShape

The *markerShape* keyword in the JSON that defines the marker shape in the shapes section needs one of the attributes shown in Table 8-13.

Table 8-13. *Marker Shape Options*

Value	Formatting Popup Option
Circle	● (circle)
Square	■ (square)
Diamond	◆ (diamond)
Triangle	(triangle)

Tooltip options are described in Table 6-5.
Legend keywords are given in Table 7-2.
Categories keywords are given in Table 7-3.
Values keywords are given in Table 7-4.
Zoom keywords are given in Table 7-5. For a column chart, the showOnValueAxis refers to the Y axis.
Labels keywords are given in Table 7-6.
Totals keywords are given in Table 7-7.
Plot area keywords are given in Table 7-8.
Small multiples options are described in Table 7-9.
Small multiples titles options are described in Table 7-10.
Legend position options are described in Table 7-11.
Category axis position options are described in Table 7-12.
Value axis display units options are described in Table 7-13.
Category axis display units options are described in Table 7-13.
Label position options are described in Table 7-14.
Gridline options are described in Table 7-15.
Value axis gridline style options are described in Table 7-16.
The color border section only contains the *show* keyword.
You can find the JSON code for this in the sample file Attributes_ScatterChart.json.

Conclusion

This chapter introduced to you six chart types: 100% stacked bar chart, 100% stacked column chart, stacked area chart, line and stacked column chart, line and clustered column chart, and scatter chart.

They are all built on chart styles (and JSON code) that you saw in the previous chapter and extend your knowledge to enable you to create theme files that can define chart formatting for all the classic data-driven chart types.

To conclude your tour of JSON formatting for charts, the next chapter will explore the remaining Power BI chart types that you can add to your theme files.

CHAPTER 9

■ ■ ■

Other Chart Visual Styles

To complete your tour of theme file chart formatting, you need to examine the JSON for a final set of chart types. These are

- Pie
- Donut
- Treemap
- Waterfall
- Funnel
- Ribbon chart

All of these are less complex than some of the charts that you have seen so far – considerably less so in some cases. Nevertheless, there are some new keywords that you will discover in this chapter that are specific to the JSON that describes the formatting of these chart types.

All the chart types discussed in this chapter have a separate JSON file in the samples folder that is available for download. Each file is self-contained and can be used to format the relevant chart type.

Pie Chart

This is certainly one of the simpler charts – and consequently maps to a correspondingly simple JSON theme file.

The keyword that specifies the formatting attributes of a pie chart is *pieChart*.

Table 9-1 outlines the JSON keywords that you can use in pie chart formatting and the section that corresponds to each keyword in the Power BI Desktop format pane.

Table 9-1. *Pie Chart Keywords*

Keyword	Formatting Section
legend	Legend
labels	Detail labels
slices	Rotation
padding	Properties/Padding
title	Title
subTitle	Subtitle
divider	Divider
spacing	Spacing
background	Background
lockAspect	Lock aspect
border	Border
dropShadow	Shadow
visualHeader	Header Icons
visualHeaderTooltip	Tooltips
visualTooltip	Tooltip

You can find the JSON code for this in the sample file Attributes_PieChart.json.

The following code snippet provides the JSON code used to pre-format a pie chart:

```
"pieChart":
  {"*":
    {
    "general": [{... code not shown ...}]
    "legend": [{... code not shown ...}],
     "labels": [{
            "show": true,
            "labelStyle": "Data",
            "color": {"solid": {"color": "#000000"}},
            "labelDisplayUnits": 1000,
            "labelPrecision": 1,
            "percentageLabelPrecision": 1,
            "fontSize": 8,
            "fontFamily": "Arial",
            "bold": true,
            "italic": false,
            "underline": false,
            "position": "preferInside",
            "overflow": true,
            "background": "off"
            }],
```

```
    "slices": [{
      "startAngle": 90
           }],
  "title": [{... code not shown ...}],
  "subTitle": [{... code not shown ...}],
  "divider": [{... code not shown ...}],
  "spacing": [{... code not shown ...}],
  "background": [{... code not shown ...}],
  "lockAspect": [{... code not shown ...}],
  "border": [{... code not shown ...}],
  "dropShadow": [{... code not shown ...}],
  "visualTooltip": [{... code not shown ...}],
  "visualHeader": [{... code not shown ...}]
  "visualHeaderTooltip": [{... code not shown ...}]
  }
}
```

Tooltip options are described in Table 6-5.

Legend keywords are given in Table 7-2.

The *labels* keyword that describes the detail labels section uses the keywords shown in Table 9-2.

Table 9-2. *Element Keyword Mapping for the Detail Labels Section*

Element	Keyword
Detail labels	show
Label style	labelStyle
Color	color
Display units	labelDisplayUnits
Value decimal places	labelPrecision
% decimal places	percentageLabelPrecision
Text size	fontSize
Font family	fontFamily
(Bold)	bold
(Italic)	italic
(Underline)	underline
Label position	position
Overflow text	overflow
Background	background

The *labelStyle* keyword in the JSON that defines label style in a pie or a donut chart needs one of the attributes shown in Table 9-3.

Table 9-3. *Label Style Options*

Value	Formatting Popup Option
Category	Category
DataValue	Data value
Percent of total	Percent of total
Category, data value	Category, data value
Category, percent of total	Category, percent of total
Data value, percent of total	Data value, percent of total
All detail labels	All detail labels

■ **Note** Taking an approach that is slightly unusual in theme files, the values used to specify label style options contain spaces and punctuation.

The *position* keyword in the JSON that defines label position in a pie or a donut chart needs one of the attributes shown in Table 9-4.

Table 9-4. *Label Position Options*

Value	Formatting Popup Option
outside	Outside
inside	Inside
preferOutside	Prefer outside
preferInside	Prefer inside

■ **Note** Interestingly, the values used to specify label position options have to be in camel case.

The *background* keyword in the JSON that defines the background in a pie or a donut chart (in the labels section) needs one of the attributes shown in Table 9-5.

Table 9-5. *Label Style Options*

Value	Formatting Popup Option
on	On
off	Off

Tooltip options are described in Table 6-5.
Legend position options are described in Table 7-11.

Donut Chart

The donut chart is virtually identical to the pie chart as far as formatting is concerned. It has one additional feature, however. This is the *inner radius* (defined by the *slices* keyword) that you can see in the JSON that is given in the following.

The keyword that specifies the formatting attributes of a donut chart is *donutChart*.

Table 9-6 outlines the JSON keywords that you can use in donut chart formatting and the section that corresponds to each keyword in the Power BI Desktop format pane.

Table 9-6. *Donut Chart Keywords*

Keyword	Formatting Section
general	General
legend	Legend
slices	Shapes
title	Title
subTitle	Subtitle
divider	Divider
spacing	Spacing
background	Background
lockAspect	Lock aspect
border	Border
dropShadow	Shadow
visualHeader	Header Icons
visualHeaderTooltip	Tooltips
visualTooltip	Tooltip

The following code snippet provides the JSON code used to pre-format a donut chart:

```
"donutChart":
  {"*":
    {
    "general": [{... code not shown ...}]
    "legend": [{... code not shown ...}],
    "labels": [{
            "show": true,
            "labelStyle": "Category, data value, percent of total",
            "color": {"solid": {"color": "#000000"}},
            "labelDisplayUnits": 1000,
            "labelPrecision": 1,
            "fontFamily": "Arial",
            "bold": true,
            "italic": false,
```

```
            "underline": false,
            "fontSize": 8,
            "position": "preferInside"
            }],
     "slices": [{
            "innerRadiusRatio": 20,
            "startAngle": 90
            }],
     "title": [{... code not shown ...}],
     "subTitle": [{... code not shown ...}],
     "divider": [{... code not shown ...}],
     "spacing": [{... code not shown ...}],
     "background": [{... code not shown ...}],
     "lockAspect": [{... code not shown ...}],
     "border": [{... code not shown ...}],
     "dropShadow": [{... code not shown ...}],
     "visualTooltip": [{... code not shown ...}],
     "visualHeader": [{... code not shown ...}]
     }
}
```

Legend keywords are given in Table 7-2.

The *slices* keyword that is used in the shapes section uses the keywords shown in Table 9-7.

Table 9-7. *Element Keyword Mapping for the Shapes Section*

Element	Keyword
Inner radius	innerRadiusRatio
Rotation	startAngle

Tooltip options are described in Table 6-5.

Legend keywords are given in Table 7-2.

Legend position options are described in Table 7-11.

You can find the complete definition of a donut chart JSON formatting in the file Attributes_Donut.json.

Treemap

As one of the simpler visuals (in terms of presentation attributes, at least), a treemap only contains three elements apart from those shared by all visuals. These are

- Legend

- Data labels

- Category labels

Of these, the latter two are the ones that contain elements specific to the treemap.

The keyword that specifies the formatting attributes of a treemap is *treemap*.

Table 9-8 outlines the JSON keywords that you can use in treemap chart formatting.

Table 9-8. *Treemap Keywords*

Keyword	Formatting Section
General	General
Legend	Legend
Labels	Data labels
categoryLabels	Category labels
Title	Title
Subtitle	Subtitle
Divider	Divider
Spacing	Spacing
background	Background
lockAspect	Lock aspect
Border	Border
dropShadow	Shadow
visualHeader	Header Icons
visualHeaderTooltip	Tooltips
visualTooltip	Tooltip

You can see this in the following code snippet that is taken from the sample file Attributes_Treemap.json:

```
"treemap":
  {"*":
    {
    "general": [{... code not shown ...}]
    "legend": [{... code not shown ...}],
      "labels": [{
            "show": true,
            "color": {"solid": {"color": "#000000"}},
            "labelDisplayUnits": 1000,
            "labelPrecision": 1,
            "fontFamily": "Arial",
            "fontSize": 10,
            "bold": true,
            "italic": false,
            "underline": false
            }],
      "categoryLabels": [{
            "show": true,
            "color": {"solid": {"color": "#ffffff"}},
            "fontFamily": "Arial",
            "fontSize": 9,
```

```
                    "bold": true,
                    "italic": false,
                    "underline": false
                    }],
            "title": [{... code not shown ...}],
            "subTitle": [{... code not shown ...}],
            "divider": [{... code not shown ...}],
            "spacing": [{... code not shown ...}],
            "background": [{... code not shown ...}],
            "lockAspect": [{... code not shown ...}],
            "border": [{... code not shown ...}],
            "dropShadow": [{... code not shown ...}],
            "visualTooltip": [{... code not shown ...}],
            "visualHeader": [{... code not shown ...}]
            "visualHeaderTooltip": [{... code not shown ...}]
            }
    }
```

Legend keywords are given in Table 7-2.

The *labels* keyword that describes the data labels section uses the keywords shown in Table 9-9.

Table 9-9. *Element Keyword Mapping for the Data Labels Section*

Element	Keyword
Data labels	show
Color	color
Display units	labelDisplayUnits
Value decimal places	labelPrecision
Text size	fontFamily
Font family	fontSize
(Bold)	bold
(Italic)	italic
(Underline)	underline

The *categoryLabels* keyword that describes the category labels section uses the keywords shown in Table 9-10.

Table 9-10. *Element Keyword Mapping for the Category Labels Section*

Element	Keyword
Category labels	Show
Color	Color
Text size	fontFamily
Font family	fontSize
(Bold)	Bold
(Italic)	Italic
(Underline)	Underline

Tooltip options are described in Table 6-5.

Label display units options are described in Table 7-11.

Waterfall Chart

One chart that has a couple of interesting new formatting elements is the waterfall chart. Specifically, it contains the sentiment colors section that you need to be able to configure.

The keyword that specifies the formatting attributes of a waterfall chart is *waterfallChart*.

Table 9-11 outlines the JSON keywords that you can use in a waterfall chart and the section that corresponds to each keyword in the Power BI Desktop format pane.

Table 9-11. *Waterfall Chart Keywords*

Keyword	Formatting Section
general	General
legend	Legend
categoryAxis	X axis
valueAxis	Y axis
labels	Data labels
plotArea	Plot area
sentimentColors	Sentiment colors
title	Title
subTitle	Subtitle
divider	Divider
spacing	Spacing
background	Background
lockAspect	Lock aspect
border	Border
dropShadow	Shadow
visualHeader	Header Icons
visualHeaderTooltip	Tooltips
visualTooltip	Tooltip

The following code snippet provides the JSON code used to pre-format a waterfall chart:

```
"waterfallChart":
  {"*":
    {
    "general": [{... code not shown ...}]
    "legend": [{... code not shown ...}],
    "categoryAxis": [{
        "show": true,
        "color": {"solid": {"color": "#949494"}},
        "fontSize": 8,
        "fontFamily": "Arial",
        "bold": true,
        "italic": false,
        "underline": false,
        "preferredCategoryWidth": 20,
        "maxMarginFactor": 25,
        "innerPadding": 15,
        "showAxisTitle": true,
```

```
        "axisStyle": "showTitleOnly",
        "titleColor": {"solid": {"color": "#949494"}},
        "titleText": "Add an Axis Title",
        "titleFontSize": 8,
        "titleFontFamily": "Arial",
        "titleBold": true,
        "titleItalic": false,
        "titleUnderline": false
        }],
"valueAxis": [{
        "show": true,
        "position": "Left",
        "start": 0,
        "end": 0,
        "labelColor": {"solid": {"color": "#949494"}},
        "fontSize": 8,
        "fontFamily": "Arial",
        "labelDisplayUnits": 1000,
        "labelPrecision": 2,
        "showAxisTitle": true,
        "axisStyle": "showTitleOnly",
        "color": {"solid": {"color": "#949494"}},
        "titleText": "Add an Axis Title",
        "titleFontSize": 8,
        "titleFontFamily": "Arial",
        "titleBold": true,
        "titleItalic": false,
        "titleUnderline": false,
        "bold": true,
        "italic": false,
        "underline": false,
        "gridLineShow": false,
        "gridLineColor":  {"solid": {"color": "#949494"}},
        "gridLineThickness": 1,
        "gridlineStyle": "solid"
    }],
"labels": [{
        "show": true,
        "color": { "solid": { "color": "#EEEEEE"}},
        "labelDisplayUnits": 1000,
        "labelPrecision": 2,
        "labelOrientation": 0,
        "labelPosition": "OutsideEnd",
        "fontSize": 11,
        "fontFamily": "Arial",
        "bold": true,
        "italic": false,
        "underline": false,
        "enableBackground": true,
        "backgroundColor": { "solid": { "color": "#949494"}},
        "backgroundTransparency": 30
```

```
            }],
    "plotArea": [{
            "transparency": 0,
            "scaling": "Fill"
            }],
    "sentimentColors": [{
            "increaseFill": { "solid": { "color": "#949494"}},
            "decreaseFill": { "solid": { "color": "#D1CAB8"}},
            "totalFill": { "solid": { "color": "#E2EB00"}},
            "otherFill": { "solid": { "color": "#E2EB00"}}
            }],
    "title": [{... code not shown ...}],
    "subTitle": [{... code not shown ...}],
    "divider": [{... code not shown ...}],
    "spacing": [{... code not shown ...}],
    "background": [{... code not shown ...}],
    "lockAspect": [{... code not shown ...}],
    "border": [{... code not shown ...}],
    "dropShadow": [{... code not shown ...}],
    "visualTooltip": [{... code not shown ...}],
    "visualHeader": [{... code not shown ...}]
    "visualHeaderTooltip": [{... code not shown ...}]
    }
}
```

Tooltip options are described in Table 6-5.

Legend keywords are given in Table 7-2.

The *categoryAxis* keyword that describes the X axis section uses the keywords shown in Table 9-12.

Table 9-12. *Element Keyword Mapping for the X Axis Section*

Element	Keyword
X axis	show
Color	color
Text size	fontSize
Font family	fontFamily
Minimum category width	preferredCategoryWidth
Maximum size	maxMarginFactor
Inner padding	innerPadding
Title	showAxisTitle
Style	axisStyle
Title color	titleColor
Axis title	titleText
Title text size	titleFontSize
(Title) Font family	titleFontFamily
(Title) Bold	titleBold
(Title) Italic	titleItalic
(Title) Underline	titleUnderline

The *valueAxis* keyword that describes the Y axis section uses the keywords shown in Table 9-13.

Table 9-13. *Element Keyword Mapping for the Y Axis Section*

Element	Keyword
Y axis	show
Position	position
Start	start
End	end
Color	color
Text size	fontSize
Font family	fontFamily
(Bold)	bold
(Italic)	italic
(Underline)	underline
Display units	labelDisplayUnits
Value decimal places	labelPrecision
Title	showAxisTitle
Style	axisStyle
Title color	labelColor
Axis title	titleText
Title text size	titleFontSize
(Title) Font family	titleFontFamily
(Title) Bold	titleBold
(Title) Italic	titleItalic
(Title) Underline	titleUnderline
Gridlines	gridLineShow
(Gridlines) Color	gridLineColor
Stroke width	gridLineThickness
Line style	gridlineStyle

The *labels* keyword that describes the labels section uses the keywords shown in Table 9-14.

Table 9-14. *Element Keyword Mapping for the Labels Section*

Element	Keyword
Data labels	show
Color	color
Display units	labelDisplayUnits
Value decimal places	labelPrecision
Orientation	labelOrientation
Position	labelPosition
Text size	fontSize
Font family	fontFamily
(Bold)	bold
(Italic)	italic
(Underline)	underline
Show background	enableBackground
Background color	backgroundColor
Transparency	backgroundTransparency

Categories keywords are given in Table 7-3.
Values keywords are given in Table 7-4.
Labels keywords are given in Table 7-6.
Totals keywords are given in Table 7-7.
Plot area keywords are given in Table 7-8.
The *sentimentColors* keyword that describes the sentiment colors section needs one of the keywords shown in Table 9-15.

Table 9-15. *Element Keyword Mapping for the Sentiment Colors Section*

Element	Keyword
Increase	increaseFill
Decrease	decreaseFill
Total	totalFill
Other	otherFill

You can find the JSON code for this in the sample file Attributes_Waterfall.json.
The sentiment colors are simple and clear – you just define the color required for each aspect of the waterfall chart that you would otherwise format directly in Power BI Desktop.
Tooltip keywords are given in Table 6-5.
Legend position options are described in Table 7-11.
Category axis position options are described in Table 7-12.

191

Value axis display units options are described in Table 7-13.
Value axis style options are described in Table 7-19.
Value axis gridline style options are described in Table 7-16.
Label position options are described in Table 7-14.

Funnel Chart

The penultimate chart that we will look at in this chapter is the funnel chart. This chart type adds only one new section keyword: *percentBarLabel*.

The keyword that specifies the formatting attributes of a funnel chart is *funnel*.

Table 9-16 outlines the JSON keywords that you can use in funnel chart formatting and the section that corresponds to each keyword in the Power BI Desktop format pane.

Table 9-16. *Funnel Chart Keywords*

Keyword	Formatting Section
General	General
categoryAxis	Category labels
dataPoint	Data colors
Labels	Data labels
percentBarLabel	Conversion rate labels
title	Title
subTitle	Subtitle
divider	Divider
spacing	Spacing
background	Background
lockAspect	Lock aspect
border	Border
dropShadow	Shadow
visualHeader	Header Icons
visualHeaderTooltip	Tooltips
visualTooltip	Tooltip

The following code snippet provides the JSON code used to pre-format a funnel chart:

```
"funnel":
 {"*":
   {
   "general": [{... code not shown ...}]
   "legend": [{... code not shown ...}],
   "categoryAxis": [{
```

```
            "show": true,
            "color": {"solid": {"color": "#949494"}},
            "fontSize": 8,
            "fontFamily": "Arial",
            "bold": true,
            "italic": false,
            "underline": false
            }],
    "dataPoint": [{
            "showAllDataPoints": true,
            "defaultColor": { "solid": { "color": "#12239E"}}
    }],
    "labels": [{
            "show": true,
            "style": "PercentFirst",
            "color": {"solid": {"color": "#ffffff"}},
            "labelDisplayUnits": 1,
            "labelPrecision": 2,
            "labelPosition": "InsideCenter",
            "fontFamily": "Arial",
            "fontSize": 7,
            "position": "OutsideEnd",
            "bold": true,
            "italic": false,
            "underline": false
            }],
    "percentBarLabel": [{
            "show": true,
            "color": {"solid": {"color": "#12239E"}},
            "fontFamily": "Arial",
            "fontSize": 16,
            "bold": true,
            "italic": false,
            "underline": false
            }],
    "title": [{... code not shown ...}],
    "subTitle": [{... code not shown ...}],
    "divider": [{... code not shown ...}],
    "spacing": [{... code not shown ...}],
    "background": [{... code not shown ...}],
    "lockAspect": [{... code not shown ...}],
    "border": [{... code not shown ...}],
    "dropShadow": [{... code not shown ...}],
    "visualTooltip": [{... code not shown ...}],
    "visualHeader": [{... code not shown ...}]
    "visualHeaderTooltip": [{... code not shown ...}]
    }
}
```

The *categoryAxis* keyword that describes the category labels section uses the keywords shown in Table 9-17.

Table 9-17. *Element Keyword Mapping for the Category Labels Section*

Element	Keyword
Category labels	Show
Color	Color
Text size	fontSize
Font family	fontFamily
(Bold)	Bold
(Italic)	Italic
(Underline)	Underline

The *dataPoint* keyword that describes the data colors section uses the keywords shown in Table 9-18.

Table 9-18. *Element Keyword Mapping for the Data Colors Section*

Element	Keyword
Show all	showAllDataPoints
Default color	defaultColor

The *labels* keyword that describes the data labels section uses the keywords shown in Table 9-19.

Table 9-19. *Element Keyword Mapping for the Data Labels Section*

Element	Keyword
Data labels	Show
Label style	Style
Color	Color
Display units	labelDisplayUnits
Value decimal places	labelPrecision
Position	labelPosition
Text size	fontSize
Font family	fontFamily
(Bold)	Bold
(Italic)	Italic
(Underline)	Underline

The *percentBarLabel* keyword that describes the Conversion Rate Label section uses the keywords shown in Table 9-20.

Table 9-20. *Element Keyword Mapping for the Conversion Rate Label Section*

Element	Keyword
Conversion Rate Label	Show
Color	Color
Text size	fontSize
Font family	fontFamily
(Bold)	Bold
(Italic)	Italic
(Underline)	Underline

The *style* keyword in the JSON that defines label position in a funnel chart needs one of the attributes shown in Table 9-21.

Table 9-21. *Label Style Options*

Value	Formatting Popup Option
Data	Data Value
Percent of first	Percent of first
Percent of previous	Percent of previous
Data, percent of first	Data value, percent of first
Data, percent of previous	Data value, percent of previous

The *position* keyword in the JSON that defines label style in a funnel chart needs one of the attributes shown in Table 9-22.

Table 9-22. *Label Position Options*

Value	Formatting Popup Option
OutsideEnd	Outside End
InsideCenter	Inside Center

Tooltip options are described in Table 6-5.
Legend keywords are given in Table 7-2.
You can find the JSON code for this in the sample file Attributes_FunnelChart.json.

Ribbon Chart

To conclude not only the chapter but also the review of chart type formatting in theme files, you need to look at the ribbon chart. This chart type is very similar to a line chart – at least as far as the JSON is concerned. The principal difference is the addition of the *ribbonChart* keyword that introduces the section on the specifics of ribbon chart formatting. Confusingly, this is the same keyword that is used to define the actual ribbon chart object in the JSON.

Table 9-23 outlines the JSON keywords that you can use in ribbon chart formatting and the section that corresponds to each keyword in the Power BI Desktop format pane.

Table 9-23. *Ribbon Chart Keywords*

Keyword	Formatting Section
general	General
legend	Legend
categoryAxis	X axis
smallMultiplesLayout	Small Multiples
subheader	Small Multiples/Title
zoom	Zoom slider
labels	Data labels
plotArea	Plot area background
ribbonChart	Ribbons
title	Title
subTitle	Subtitle
divider	Divider
spacing	Spacing
background	Background
lockAspect	Lock aspect
border	Border
dropShadow	Shadow
visualHeader	Header Icons
visualHeaderTooltip	Tooltips
visualTooltip	Tooltip

The following code snippet provides the JSON code used to pre-format a ribbon chart:

```
"ribbonChart":
  {"*":
    {
    "general": [{... code not shown ...}]
    "legend": [{... code not shown ...}],
```

```
"categoryAxis": [{
    "show": true,
    "color": {"solid": {"color": "#949494"}},
    "fontSize": 8,
    "fontFamily": "Arial",
    "bold": true,
    "italic": false,
    "underline": false,
    "preferredCategoryWidth": 20,
    "maxMarginFactor": 30,
    "concatenateLabels": true,
    "showAxisTitle": true,
    "axisStyle": "showTitleOnly",
    "titleColor": {"solid": {"color": "#949494"}},
    "titleText": "Add an Axis Title",
    "titleFontSize": 8,
    "titleFontFamily": "Arial",
    "concatenateLabels": true
    }],
"ribbonChart": [{
    "seriesGapRatio": 5,
    "colorBands": true,
    "bandsTransparency": 50,
    "showBorder": true
      }],
"labels": [{
    "show": true,
    "color": {"solid": {"color": "#ffffff"}},
    "labelDisplayUnits": 1,
    "labelPrecision": 2,
    "labelPosition": "InsideEnd",
    "labelOverflow": true,
    "labelOrientation": 0,
    "fontFamily": "Arial",
    "fontSize": 7,
    "bold": true,
    "italic": false,
    "underline": false,
    "enableBackground": false,
    "backgroundColor": {"solid": {"color": "#949494"}},
    "backgroundTransparency": 0
      }],
"plotArea": [{... code not shown ...}],
"title": [{... code not shown ...}],
"subTitle": [{... code not shown ...}],
"divider": [{... code not shown ...}],
"spacing": [{... code not shown ...}],
"background": [{... code not shown ...}],
"lockAspect": [{... code not shown ...}],
"border": [{... code not shown ...}],
"dropShadow": [{... code not shown ...}],
```

```
    "visualTooltip": [{... code not shown ...}],
    "visualHeader": [{... code not shown ...}]
    "visualHeaderTooltip": [{... code not shown ...}]
  }
```

Legend keywords are given in Table 7-2.

The *categoryAxis* keyword that describes the X axis section uses the keywords shown in Table 9-24.

Table 9-24. *Element Keyword Mapping for the X Axis Section*

Element	Keyword
X axis	Show
Color	Color
Text size	fontSize
Font family	fontFamily
Minimum category width	preferredCategoryWidth
Maximum size	maxMarginFactor
Concatenate labels	concatenateLabels
Title	showAxisTitle
Style	axisStyle
Title color	titleColor
Axis title	titleText
Title text size	titleFontSize
(Axis) Font family	titleFontFamily
Concatenate labels	concatenateLabels

Zoom keywords are given in Table 7-5.

The *labels* keyword that describes the data labels section uses the keywords shown in Table 9-25.

Table 9-25. *Element Keyword Mapping for the Data Labels Section*

Element	Keyword
Data labels	Show
Color	Color
Display units	labelDisplayUnits
Value decimal places	labelPrecision
Orientation	labelOrientation
Position	labelPosition
Overflow text	labelOverflow
Text size	fontSize
Font family	fontFamily
Show background	enableBackground
Background color	backgroundColor
Transparency	backgroundTransparency

Legend keywords are given in Table 7-2.
Tooltip options are described in Table 6-5.
Plot area keywords are given in Table 7-8.
The *ribbonChart* keyword that describes the ribbons section uses the keywords shown in Table 9-26.

Table 9-26. *Element Keyword Mapping for the Ribbons Section*

Element	Keyword
Ribbons	Show
Spacing	seriesGapRatio
Match series color	colorBands
Transparency	bandsTransparency
Border	showBorder

Categories keywords are given in Table 7-3.
Labels keywords are given in Table 7-6.
Plot area keywords are given in Table 7-8.
You can find the JSON code for this in the sample file Attributes_RibbonChart.json.

Conclusion

This chapter introduced you to the JSON code that you need to format the six remaining chart types. You saw the specifics of preparing the JSON for pie, donut, treemap, waterfall, funnel, and ribbon charts.

Now that the three-chapter tour of chart types is finished, it is time to move on to another completely different category of visuals in Power BI – maps. These are the subject of the next chapter.

CHAPTER 10

■ ■ ■

Maps

This chapter looks at a completely different type of visual to those covered in the preceding three chapters. Here, we will look at how maps can be pre-formatted using theme files. Here, you will see how to prepare the JSON to define the presentation of the following visual types:

- Map
- Filled map
- ArcGIS map

Each of the map types discussed in this chapter has a separate JSON file in the samples folder that contains the code that you can use to define your own theme files.

Map

The presentation aspects of a Power BI map that you can format (apart from the standard elements that are common to nearly all visuals) are

- Legend
- Data colors
- Category
- Bubbles
- Map controls
- Map styles
- Heat map

The keyword that specifies the formatting attributes of a map is *map*.

Table 10-1 outlines the JSON keywords that you can use in map formatting and their corresponding section in the Power BI Desktop format pane.

© Adam Aspin 2023
A. Aspin, *Pro Power BI Theme Creation*, https://doi.org/10.1007/978-1-4842-9633-2_10

Table 10-1. *Map Chart Keywords*

Keyword	Formatting Section
General	General
Legend	Legend
dataPoint	Data colors
categoryLabels	Category
Bubbles	Bubbles
mapControls	Map controls
mapStyles	Map styles
Title	Title/Title
subtitle	Title/Subtitle
Divider	Title/Divider
Spacing	Title/Spacing
Background	Effects/Background
Border	Effects/Visual border
dropShadow	Effects/Shadow
visualHeader	Header Icons
lockAspect	Lock aspect
visualHeaderTooltip	Header Icons/Help tooltip
visualTooltip	Tooltips

You can find the JSON code for this in the sample file Attributes_Map.json.

The following code snippet provides the JSON code used to pre-format a map. As maps are a new concept, I have preferred to show virtually all the JSON code:

```
"map":{"*":
  {
    "general": [{... code not shown ...}]
    "legend": [{... code not shown ...}],
     "mapStyles": [{
            "mapTheme": "grayscale",
            "showLabels": true
            }],
     "mapControls": [{
            "autoZoom": true,
            "showZoomButtons": true,
            "showLassoButton": true,
            "geocodingCulture": ""
            }],
```

```
    "legend": [{
        "show": false,
        "position": "Top",
        "showTitle": true,
        "titleText": "Add a legend title",
        "legendColor": {"solid": {"color": "#949494"}},
        "fontFamily": "Arial",
        "fontSize": 10,
        "bold": true,
        "italic": false,
        "underline": false
        }],
    "categoryLabels": [{
        "show": true,
        "color": {"solid": {"color": "#949494"}},
        "fontSize": 10,
        "fontFamily": "Arial",
        "enableBackground": true,
        "backgroundColor": {"solid": {"color": "#ffffff"}},
        "transparency": 30
        }],
    "bubbles": [{
        "bubbleSize": 15
        }],
    "heatMap":    [{
        "show": true,
        "filterRadius": 8,
        "transparency": 10,
        "unit": "meters",
        "color": {"solid": {"color": "#F5F4F0"}},
        "color50": {"solid": {"color": "#949494"}},
        "color100": {"solid": {"color": "#D5D5D5"}}
        }],
    "title": [{... code not shown ...}],
    "subTitle": [{... code not shown ...}],
    "divider": [{... code not shown ...}],
    "spacing": [{... code not shown ...}],
    "background": [{... code not shown ...}],
    "lockAspect": [{... code not shown ...}],
    "border": [{... code not shown ...}],
    "dropShadow": [{... code not shown ...}],
    "visualHeader": [{... code not shown ...}]
    "visualHeaderTooltip": [{... code not shown ...}]
    "visualTooltip": [{... code not shown ...}]
    "padding": [{... code not shown ...}]
  }
}
```

Legend keywords are given in Table 7-2.

The *categoryLabels* keyword that describes the category section uses the keywords shown in Table 10-2.

Table 10-2. *Keyword Mapping for the Category Labels Section*

Element	Keyword
Category labels	show
Color	color
Text size	fontSize
Font family	fontFamily
Show background	enableBackground
Background color	backgroundColor
Transparency	transparency

The *bubbles* keyword that describes the bubbles section uses the keyword shown in Table 10-3.

Table 10-3. *Keyword Mapping for the Bubbles Section*

Element	Keyword
Size	bubbleSize

The *mapControls* keyword that describes the map controls section uses the keywords shown in Table 10-4.

Table 10-4. *Keyword Mapping for the Map Controls Section*

Element	Keyword
Map controls	Show
Auto zoom	autoZoom
Zoom buttons	showZoomButtons

The *mapStyles* keyword that describes the map styles section uses the keywords shown in Table 10-5.

Table 10-5. *Keyword Mapping for the Map Styles Section*

Element	Keyword
Theme	mapTheme
Show labels	showLabels

The *heatMap* keyword that describes the heat map section uses the keywords shown in Table 10-6.

Table 10-6. *Keyword Mapping for the Heat Map Section*

Element	Keyword
Heat map	Show
Radius	filterRadius
Transparency	transparency
Unit	Units
0% gradient stop	Color
50% gradient stop	Color50
100% gradient stop	Color100

The *mapTheme* keyword in the map styles section requires one of the attributes shown in Table 10-7.

Table 10-7. *Map Styles Options*

Value	Formatting Popup Option
aerial	Aerial
dark	Dark
light	Light
grayscale	Grayscale
road	Road

The *unit* keyword for heat maps (in the heat map section) needs one of the attributes shown in Table 10-8.

Table 10-8. *Units Options*

Value	Formatting Popup Option
pixels	Pixels
meters	Meters

Tooltip options are described in Table 6-5.
Legend position options are described in Table 7-11.

Filled Map

The JSON code that is required to format a filled map is very similar to that used for a map.

The keyword that specifies the formatting attributes of a filled map is *filledMap*.

Table 10-9 outlines the JSON keywords that you can use in filled map formatting and their corresponding section in the Power BI Desktop format pane.

Table 10-9. *Filled Map Chart Keywords*

Keyword	Formatting Section
General	General
dataPoint	Data colors
Shape	Shape
Zoom	Zoom
Title	Title/Title
Subtitle	Title/Subtitle
Divider	Title/Divider
Spacing	Title/Spacing
Background	Effects/Background
Border	Effects/Visual border
dropShadow	Effects/Shadow
visualHeader	Header Icons/Icons Header Icons/Colors
lockAspect	Lock aspect
visualHeaderTooltip	Header Icons/Help tooltip
visualTooltip	Tooltips

You can find the JSON code for this in the sample file Attributes_FilledMap.json.
The following code snippet provides the JSON code used to pre-format a filled map:

```
"filledMap":
  {"*":
    {
    "general": [{... code not shown ...}]
      "mapStyles": [{
            "mapTheme": "grayscale",
            "showLabels": true
            }],
      "mapControls": [{
            "autoZoom": true,
            "showZoomButtons": true,
            "showLassoButton": true,
            "geocodingCulture": ""
            }],
    "title": [{... code not shown ...}],
    "subTitle": [{... code not shown ...}],
    "divider": [{... code not shown ...}],
    "spacing": [{... code not shown ...}],
    "background": [{... code not shown ...}],
    "lockAspect": [{... code not shown ...}],
```

```
"border": [{... code not shown ...}],
"dropShadow": [{... code not shown ...}],
"visualHeader": [{... code not shown ...}]
"visualHeaderTooltip": [{... code not shown ...}]
"visualTooltip": [{... code not shown ...}]
"padding": [{... code not shown ...}]
}
}
```

Map controls elements are described in Table 10-4.
Map style elements are described in Table 10-7.
Tooltip options are described in Table 6-5.
Legend position options are described in Table 7-11.

ArcGIS Map

Table 10-10 describes the JSON keywords that correspond to each formatting section for ArcGIS maps in Power BI Desktop. This uses the keyword *webMapExtensions*.

Table 10-10. *ArcGIS Maps Formatting Elements*

Keyword	Formatting Section
general	General
title	Title
subTitle	Subtitle
divider	Divider
spacing	Spacing
background	Background
lockAspect	Lock aspect
border	Border
dropShadow	Shadow
visualHeader	Header icons
visualHeaderTooltip	Header icons
visualTooltip	Tooltips

The following code snippet gives you the core elements used to specify standard card formatting. The sample file Attributes_ArcGISMap.json contains all the elements that can be set for a card in a theme file.

```
"webmapExtensions":{
  "*" :
    {
       "general": [{... code not shown ...}]
       "title": [{... code not shown ...}],
       "subTitle": [{... code not shown ...}],
```

```
        "divider": [{... code not shown ...}],
        "spacing": [{... code not shown ...}],
         "background": [{... code not shown ...}],
        "lockAspect": [{... code not shown ...}],
        "border": [{... code not shown ...}],
        "dropShadow": [{... code not shown ...}],
        "visualHeader": [{... code not shown ...}]
        "visualHeaderTooltip": [{... code not shown ...}]
        "visualTooltip": [{... code not shown ...}]
        "padding": [{... code not shown ...}]
    }
}
```

Tooltip options are described in Table 6-5.

Legend position options are described in Table 7-11.

Conclusion

In this short chapter, you saw how to create the JSON for the map and filled map visuals. They use fairly simple JSON (assuming that you are used to theme file creation) but nonetheless allow you to define the essential aspects of map visual formatting.

CHAPTER 11

▪ ▪ ▪

Miscellaneous Visual Styles

There is a group of visual types that are not text based, charts, or maps. I have gathered this collection of visuals together in a single chapter to explain the JSON they use in a theme file. These visuals are

- Gauge
- Python visual
- R script visual
- Key influencers visual
- Decomposition tree visual
- KPI
- Paginated report
- Power Apps

Each of the visuals discussed in this chapter has a separate JSON file in the samples folder that contains the code that you can use to define your own theme files.

Admittedly, this collection of visuals is somewhat eclectic as they have little in common. Let's take a look at them.

Gauge

The appearance of a Power BI gauge visual can be standardized using a theme file. The keyword that specifies the formatting attributes of a gauge is *gauge*.

Table 11-1 outlines the JSON keywords that you can use in gauge formatting and the section that corresponds to each keyword in the Power BI Desktop format pane.

© Adam Aspin 2023
A. Aspin, *Pro Power BI Theme Creation*, https://doi.org/10.1007/978-1-4842-9633-2_11

Table 11-1. *Gauge Keywords*

Keyword	Formatting Section
general	General
axis	Gauge Axis
datapoint	Colors
labels	Data labels
target	Target label
calloutValue	Callout value
padding	Properties/Padding
title	Title/Title
subTitle	Title/Subtitle
divider	Title/Divider
spacing	Title/Spacing
background	Effects/Background
border	Effects/Visual border
dropShadow	Effects/Shadow
visualHeader	Header Icons
lockAspect	Lock aspect
visualHeaderTooltip	Header Icons/Help tooltip
visualTooltip	Tooltips

The following code snippet provides the JSON code used to pre-format a gauge:

```
"gauge":
 {"*":
   {
   "general": [{"responsive": true }],
    "axis": [{
          "min": 0,
          "max": 0,
          "target": 0
          }],
    "dataPoint": [{
          "fill": {"solid": {"color": "#E34B20"}},
          "target": {"solid": {"color": "#E2EB00"}}
          }],
    "labels": [{
          "show": true,
          "color": {"solid": {"color": "#615E55"}},
          "labelDisplayUnits": 0,
```

```
            "labelPrecision": 0,
            "fontSize": 8,
            "fontFamily": "Arial",
            "bold": true,
            "italic": false,
            "underline": false
            }],
    "target": [{
            "show": true,
            "color": {"solid": {"color": "#E34B20"}},
            "labelDisplayUnits": 0,
            "labelPrecision": 0,
            "fontFamily": "Arial",
            "fontSize": 8,
            "bold": true,
            "italic": false,
            "underline": false
            }],
    "calloutValue": [{
            "show": true,
            "color": {"solid": {"color": "#F5F4F0"}},
            "labelDisplayUnits": 0,
            "labelPrecision": 0,
            "fontFamily": "Arial",
            "bold": true,
            "italic": false,
            "underline": false
            }],
    "title": [{... code not shown ...}],
    "subTitle": [{... code not shown ...}],
    "divider": [{... code not shown ...}],
    "spacing": [{... code not shown ...}],
    "background": [{... code not shown ...}],
    "lockAspect": [{... code not shown ...}],
    "border": [{... code not shown ...}],
    "dropShadow": [{... code not shown ...}],
    "visualHeader": [{... code not shown ...}]
    "visualHeaderTooltip": [{... code not shown ...}]
    "visualHeader": [{... code not shown ...}]
    "padding": [{... code not shown ...}]
    }
}
```

The *axis* keyword that describes the Gauge axis section uses the keywords shown in Table 11-2.

Table 11-2. *Keyword Mapping for the Gauge Axis Section*

Element	Keyword
Min	Min
Max	Max
Target	Target

The *dataPoint* keyword that describes the data colors section uses the keywords shown in Table 11-3.

Table 11-3. *Keyword Mapping for the Data Colors Section*

Element	Keyword
Fill	Fill
Target	Target

The *labels* keyword that describes the data labels section uses the keywords shown in Table 11-4.

Table 11-4. *Keyword Mapping for the Data Labels Section*

Element	Keyword
Data labels	Show
Color	Color
Display units	labelDisplayUnits
Value decimal places	labelPrecision
Text size	fontSize
Font family	fontFamily
(Bold)	Bold
(Italic)	Italic
(Underline)	Underline

The *target* keyword that describes the target section uses the keywords shown in Table 11-5.

Table 11-5. *Keyword Mapping for the Target Section*

Element	Keyword
Target	Show
color	Color
Display units	labelDisplayUnits
Value decimal places	labelPrecision
Text size	fontSize
Font family	fontFamily
(Bold)	Bold
(Italic)	Italic
(Underline)	Underline

The *calloutValue* keyword that describes the callout value section uses the keywords shown in Table 11-6.

Table 11-6. *Keyword Mapping for the Callout Value Section*

Element	Keyword
Callout value	Show
Color	Color
Display units	labelDisplayUnits
Value decimal places	labelPrecision
Font family	fontFamily
(Bold)	Bold
(Italic)	Italic
(Underline)	Underline

Tooltip options are described in Table 6-5.
You can find the JSON code for this in the sample file Attributes_Gauge.json.

Python Visual

Python visuals are somewhat exceptional among the standard Power BI visuals inasmuch as *only* the standard visual elements (background, lock aspect, border, shadow, and the visual header) can be modified. This is because all other formatting is defined in the Python code that creates the visual.

The keyword that specifies the formatting attributes of a Python visual is *pythonVisual*.

Table 11-7 outlines the JSON keywords that you can use when formatting Python visuals and the section that corresponds to each keyword in the Power BI Desktop format pane.

Table 11-7. *Python Visual Keywords*

Keyword	Formatting Section
general	General
padding	Properties/Padding
title	Title/Title
subTitle	Title/Subtitle
divider	Title/Divider
spacing	Title/Spacing
background	Effects/Background
border	Effects/Visual border
dropShadow	Effects/Shadow
visualHeader	Header Icons
lockAspect	Lock aspect
visualHeaderTooltip	Header Icons/Help tooltip

You can find the JSON code for this in the sample file Attributes_Python.json.

The following code snippet provides the JSON code used to pre-format a Python visual. I realize that you have seen it before, but wanted to make the point that common visual elements can be specified for this visual in the theme file even if nothing else can.

```
"pythonVisual":
  {"*":
    {
    "general":  [{"responsive": true }],
    "title":    [{
        "show": true,
        "text": "Gauge Title",
        "fontColor": {"solid": {"color": "#ffffff"}},
        "background": {"solid": {"color": "#12239E"}},
        "alignment": "Center",
        "fontSize": 10,
        "fontFamily": "Arial",
          "bold": true,
          "italic": false,
          "underline": false
        }],
    "background":   [{
        "show": true,
        "color": {"solid": {"color": "#F5F4F0"}}
        }],
    "lockAspect": [{
        "show": false
          }],
```

```
"border":    [{
     "show": true,
     "color": {"solid": {"color": "#F5F4F0"}}
     }],
"subTitle": [{... code not shown ...}],
"divider": [{... code not shown ...}],
"spacing": [{... code not shown ...}],
"dropShadow": [{... code not shown ...}],
"visualHeader": [{... code not shown ...}]
"visualHeaderTooltip": [{... code not shown ...}]
"padding": [{... code not shown ...}]
}
}
```

R Script Visual

The R visual is virtually identical to the Python visual in what can be formatted. In other words, only the standard visual elements (background, lock aspect, border, shadow, and the visual header) can be modified. All other formatting is defined in the R code that creates the visual.

The keyword that specifies the formatting attributes of an R visual is *scriptVisual*.

Table 11-8 outlines the JSON keywords that you can use when formatting R visuals and the section that corresponds to each keyword in the Power BI Desktop format pane.

Table 11-8. *R Visual Keywords*

Keyword	Formatting Section
General	General
Padding	Properties/Padding
title	Title/Title
subTitle	Title/Subtitle
divider	Title/Divider
spacing	Title/Spacing
background	Effects/Background
border	Effects/Visual border
dropShadow	Effects/Shadow
visualHeader	Header Icons
lockAspect	Lock aspect
visualHeaderTooltip	Header Icons/Help tooltip

You can find the JSON code for this in the sample file Attributes_R.json.

The following code snippet provides the JSON code used to pre-format an R visual:

```
"scriptVisual":
  {"*":
    {
    "general":  [{"responsive": true }],
    "plotArea": [{... code not shown ...}],
    "title": [{... code not shown ...}],
    "subTitle": [{... code not shown ...}],
    "divider": [{... code not shown ...}],
    "spacing": [{... code not shown ...}],
    "background": [{... code not shown ...}],
    "lockAspect": [{... code not shown ...}],
    "border": [{... code not shown ...}],
    "dropShadow": [{... code not shown ...}],
    "visualHeader": [{... code not shown ...}]
    "visualHeaderTooltip": [{... code not shown ...}]
    "padding": [{... code not shown ...}]
    }
  }
```

Key Influencers Visual

The key influencers visual allows you to set three theme options that are entirely specific to this visual type. They are

- Analysis

- Analysis visual colors

- Drill visual colors

The keyword that specifies the formatting attributes of a key influencers visual is *keyDriversVisual*.

Table 11-9 outlines the JSON keywords that you can use when formatting key influencers visuals and the section that corresponds to each keyword in the Power BI Desktop format pane.

Table 11-9. *Key Influencers Visual Keywords*

Keyword	Formatting Section
General	General
keyDrivers	Analysis
keyInfluencersVisual	Analysis visual colors
keyDriversDrillVisual	Drill visual colors
padding	Properties/Padding
title	Title/Title
subTitle	Title/Subtitle
divider	Title/Divider
spacing	Title/Spacing
background	Effects/Background
border	Effects/Visual border
dropShadow	Effects/Shadow
visualHeader	Header Icons
lockAspect	Lock aspect
visualHeaderTooltip	Header Icons/Help tooltip
visualTooltip	Tooltips

The following code snippet provides the JSON code used to pre-format a key influencers visual:

```
"keyDriversVisual":
  {"*":
    {
    "general":  [{"responsive": true }],
    "keyDrivers": [{
        "allowKeyDrivers": true,
        "allowProfiles": true,
        "allowKeyDriversCounting": true,
        "countType": "relative"
        }],
    "keyInfluencersVisual": [{
        "primaryColor": {"solid": {"color": "#615E55"}},
        "primaryFontColor": {"solid": {"color": "#00AC98"}},
        "secondaryColor": {"solid": {"color": "#73268c"}},
        "secondaryFontColor": {"solid": {"color": "#465437"}},
        "canvasColor": {"solid": {"color": "#465437"}},
        "fontColor": {"solid": {"color": "#FFFFFF"}}
        }],
```

```
"keyDriversDrillVisual": [{
      "defaultColor": {"solid": {"color": "#465437"}},
      "referenceLineColor": {"solid": {"color": "#00AC98"}}
      }],
"title": [{... code not shown ...}],
"subTitle": [{... code not shown ...}],
"divider": [{... code not shown ...}],
"spacing": [{... code not shown ...}],
"background": [{... code not shown ...}],
"lockAspect": [{... code not shown ...}],
"border": [{... code not shown ...}],
"dropShadow": [{... code not shown ...}],
"visualHeader": [{... code not shown ...}]
"visualHeaderTooltip": [{... code not shown ...}]
 "visualHeader": [{... code not shown ...}]
"padding": [{... code not shown ...}]
 }
}
```

The *keyDrivers* keyword that describes the analysis section uses the keywords shown in Table 11-10.

Table 11-10. *Element Keyword Mapping for the Analysis Section*

Element	Keyword
Enable key influencers	allowKeyDrivers
Enable segments	allowProfiles
Enable counts	allowKeyDriversCounting
Count type	countType

The *keyInfluencersVisual* keyword that describes the analysis visual colors section uses the keywords shown in Table 11-11.

Table 11-11. *Keyword Mapping for the Analysis Visual Colors Section*

Element	Keyword
Primary color	primaryColor
Primary text color	primaryFontColor
Secondary color	secondaryColor
Secondary text color	secondaryFontColor
Background color	canvasColor
Font color	fontColor

The *keyDriversDrillVisual* keyword that describes the drill visual colors section uses the keywords shown in Table 11-12.

Table 11-12. *Keyword Mapping for the Drill Visual Colors Section*

Element	Keyword
Default color	defaultColor
Reference line color	referenceLineColor

The *countType* keyword in the keyDrivers section needs one of the attributes shown in Table 11-13.

Table 11-13. *Key Drivers Count Type Options*

Value	Formatting Popup Option
Absolute	Absolute
Relative	Relative

You can find the JSON code for this in the sample file Attributes_KeyInfluencers.json.

Decomposition Tree Visual

The penultimate visual that we will look at in this chapter is the decomposition tree visual. Once again, the formatting attributes that you can set in the theme file are intrinsically specific to this visual type.

The keyword that specifies the formatting attributes of a decomposition tree visual is *decompositionTreeVisual*.

Table 11-14 outlines the JSON keywords that you can use when formatting decomposition tree visuals and the section that corresponds to each keyword in the Power BI Desktop format pane.

Table 11-14. *Decomposition Tree Visual Keywords*

Keyword	Formatting Section
general	General
analysis	Analysis
tree	Tree
dataBars	Data bars
categoryLabels	Category labels
dataLabels	Data labels
levelHeader	Tree
padding	Properties/Padding
title	Title/Title
subTitle	Title/Subtitle

(continued)

Table 11-14. (*continued*)

Keyword	Formatting Section
divider	Title/Divider
spacing	Title/Spacing
background	Effects/Background
border	Effects/Visual border
dropShadow	Effects/Shadow
visualHeader	Header Icons
lockAspect	Lock aspect
visualHeaderTooltip	Header Icons/Help tooltip
visualTooltip	Tooltips

The following code snippet provides the JSON code used to pre-format a decomposition tree visual:

```
"decompositionTreeVisual":
  {"*":
  {
  "general":  [{"responsive": true }],
      "analysis": [{
            "aiEnabled": true,
            "aiMode": "relative"
      }],
      "tree": [{
            "density": "sparse",
            "accentColor": { "solid": { "color": "#F5F4F0"}},
            "connectorDefaultColor": { "solid": { "color": "#118DFF"}},
            "connectorType": "round",
            "defaultClickAction": "toggle",
            "responsiveLayout": true,
            "barsPerLevel": 10
      }],
      "dataBars": [{
            "positiveBarColor": { "solid": { "color": "#12239E"}},
            "negativeBarColor": { "solid": { "color": "#DD001E"}},
            "dataBarBackgroundColor": { "solid": { "color": "#118DFF"}},
            "dataBarWidthPercent": 100,
            "dataBarScalingType": "topNode",
            "axisStart": 0,
            "axisEnd": 0
      }],
      "categoryLabels": [{
            "categoryLabelFontFamily": "Arial",
            "categoryLabelFontSize": 12,
            "categoryLabelFontColor": { "solid": { "color": "#808080"}},
            "categoryLabelBold": true,
```

```
                "categoryLabelItalic": false,
                "categoryLabelUnderline": false
        }],
        "dataLabels": [{
                "dataLabelFontFamily": "Arial",
                "dataLabelBold": true,
                "dataLabelItalic": false,
                "dataLabelUnderline": false,
                "dataLabelFontSize": 14,
                "dataLabelFontColor": { "solid": { "color": "#808080"}},
                "dataLabelDisplayUnits": 1000,
                "dataLabelPrecision": 1
        }],
        "levelHeader": [{
                "levelHeaderBackgroundColor": { "solid": { "color": "#000000"}},
                "levelTitleFontFamily": "Arial",
                "levelTitleBold": true,
                "levelTitleItalic": false,
                "levelTitleUnderline": false,
                "levelTitleFontSize": 20,
                "levelTitleFontColor": { "solid": { "color": "#808080"}},
                "showSubtitles": false,
                "levelSubtitleFontColor": { "solid": { "color": "#FFFFFF"}},
                "levelSubtitleFontFamily": "Arial",
                "levelSubtitleBold": true,
                "levelSubtitleItalic": false,
                "levelSubtitleUnderline": false,
                "levelSubtitleFontSize": 18
        }],
    "title": [{... code not shown ...}],
    "subTitle": [{... code not shown ...}],
    "divider": [{... code not shown ...}],
    "spacing": [{... code not shown ...}],
    "background": [{... code not shown ...}],
    "lockAspect": [{... code not shown ...}],
    "border": [{... code not shown ...}],
    "dropShadow": [{... code not shown ...}],
    "visualHeader": [{... code not shown ...}]
    "visualHeaderTooltip": [{... code not shown ...}]
    "visualHeader": [{... code not shown ...}]
    "padding": [{... code not shown ...}]
    }
}
```

The *analysis* keyword that describes the analysis section uses the keywords shown in Table 11-15.

Table 11-15. *Keyword Mapping for the Analysis Section*

Element	Keyword
Enable AI splits	aiEnabled
Analysis type	aiMode

The *tree* keyword that describes the tree section uses the keywords shown in Table 11-16.

Table 11-16. *Keyword Mapping for the Tree Section*

Element	Keyword
Density	Density
Primary color	accentColor
Deselected connectors	connectorDefaultColor
Connector shape	connectorType
Default action	defaultClickAction
Responsive	responsiveLayout
Max bars shown	barsPerLevel

The *dataBars* keyword that describes the data bars section uses the keywords shown in Table 11-17.

Table 11-17. *Element Keyword Mapping for the Data Bars Section*

Element	Keyword
Positive bar	positiveBarColor
Negative bar	negativeBarColor
Bar background	dataBarBackgroundColor
Size	dataBarWidthPercent
Scale to	dataBarScalingType
Start	axisStart
End	axisEnd

The *categoryLabels* keyword that describes the category labels section uses the keywords shown in Table 11-18.

Table 11-18. *Element Keyword Mapping for the Category Labels Section*

Element	Keyword
Font family	categoryLabelFontFamily
Text size	categoryLabelFontSize
Text color	categoryLabelFontColor
(Bold)	categoryLabelBold
(Italic)	categoryLabelItalic
(Underline)	categoryLabelUnderline

The *dataLabels* keyword that describes the data labels section uses the keywords shown in Table 11-19.

Table 11-19. *Keyword Mapping for the Data Labels Section*

Element	Keyword
Font family	dataLabelFontFamily
Text size	dataLabelFontSize
(Bold)	dataLabelBold
(Italic)	dataLabelItalic
(Underline)	dataLabelUnderline
Text color	dataLabelFontColor
Display units	dataLabelDisplayUnits
Value decimal places	dataLabelPrecision

The *levelHeader* keyword that describes the level header section uses the keywords shown in Table 11-20.

Table 11-20. *Keyword Mapping for the Level Header Section*

Element	Keyword
Background color	levelHeaderBackgroundColor
Title font family	levelTitleFontFamily
Title font size	levelTitleFontSize
(Bold)	levelTitleBold
(Italic)	levelTitleItalic
(Underline)	levelTitleUnderline
Title color	levelTitleFontColor
Show subtitles	showSubtitles
Subtitle font family	levelSubtitleFontColor
Subtitle font size	levelSubtitleFontFamily
(Bold)	levelSubtitleBold
(Italic)	levelSubtitleItalic
(Underline)	levelSubtitleUnderline
Subtitle color	levelSubtitleFontSize

The *aiMode* keyword in the analysis section needs one of the attributes shown in Table 11-21.

Table 11-21. *AI Mode Options*

Value	Formatting Popup Option
absolute	Absolute
relative	Relative

The *density* keyword in the JSON for the tree needs one of the attributes shown in Table 11-22.

Table 11-22. *Density Options*

Value	Formatting Popup Option
dense	Dense
default	Default
sparse	Sparse

The *connectorType* keyword in the JSON for the tree needs one of the attributes shown in Table 11-23.

Table 11-23. *Connector Type Options*

Value	Formatting Popup Option
default	Default
round	Round

The *defaultAction* keyword in the JSON for the tree needs one of the attributes shown in Table 11-24.

Table 11-24. *Default Action Options*

Value	Formatting Popup Option
toggle	Collapse
click	Default

The *dataBarScalingType* keyword in the JSON for the data bars needs one of the attributes shown in Table 11-25.

Table 11-25. *Data Bar Scaling Type Options*

Value	Formatting Popup Option
topNode	Top node
parentNode	Parent node
levelMaximum	Level maximum

Conditional formatting is, again, not possible to define in the JSON theme file.

Data label display units options are shown in Table 7-13.

You can find the JSON code for this in the sample file Attributes_DecompositionTree.json.

KPI

The next piece of JSON code that you will see in this chapter is for the KPI visual. It is introduced by the *kpi* keyword in a theme file.

Table 11-26 outlines the JSON keywords that you can use when formatting decomposition tree visuals and the section that corresponds to each keyword in the Power BI Desktop format pane.

Table 11-26. *KPI Visual Keywords*

Keyword	Formatting Section
general	General
indicator	Callout value, Icons
trendline	Trend axis
goals	Target label
status	Trend axis
padding	Properties/Padding
title	Title/Title
subTitle	Title/Subtitle
divider	Title/Divider
spacing	Title/Spacing
background	Effects/Background
border	Effects/Visual border
dropShadow	Effects/Shadow
visualHeader	Header Icons
lockAspect	Lock aspect
visualHeaderTooltip	Header Icons/Help tooltip
visualTooltip	Tooltips

The code to format a KPI is as follows:

```
"kpi":
 {"*" :
   {
   "general":  [{"responsive": true }],
    "indicator":    [{
           "indicatorDisplayUnits": 0,
           "indicatorPrecision": 2,
           "fontSize": 20,
           "fontColor": {"solid": {"color": "#F5F4F0"}},
           "fontFamily": "Arial",
           "horizontalAlignment": "center",
           "verticalAlignment": "middle",
           "showIcon": true,
           "iconSize": 15,
           "bold": true,
           "italic": false,
           "underline": false
           }],
      "trendline":  [{
           "show": true,
           "transparency": 30
           }],
      "goals": [{
           "showGoal": true,
           "goalText": "Target",
           "showDistance": true,
           "goalFontColor": {"solid": {"color": "#000000"}},
           "goalFontFamily": "Arial",
           "fontSize": 9,
           "showDistance": true,
           "distanceLabel": "Value",
           "labelPrecision": 2,
           "distanceFontFamily": "Arial",
           "distanceFontColor": {"solid": {"color": "#FFFFFF"}},
           "titleFontSize": 13,
           "direction": "Low is good",
           "bold": true,
           "italic": false,
           "underline": false
           }],
      "status": [{
           "direction": "Positive",
           "goodColor": {"solid": {"color": "#F5F4F0"}},
           "neutralColor": {"solid": {"color": "#99AC98"}},
           "badColor": {"solid": {"color": "#FF001E"}}
           }],
    "title": [{... code not shown ...}],
    "subTitle": [{... code not shown ...}],
    "divider": [{... code not shown ...}],
```

```
    "spacing": [{... code not shown ...}],
    "background": [{... code not shown ...}],
    "lockAspect": [{... code not shown ...}],
    "border": [{... code not shown ...}],
    "dropShadow": [{... code not shown ...}],
    "visualHeader": [{... code not shown ...}]
    "visualHeaderTooltip": [{... code not shown ...}]
    "visualHeader": [{... code not shown ...}]
    "padding": [{... code not shown ...}]
    }
}
```

The *indicator* keyword that describes the indicator section uses the keywords shown in Table 11-27.

Table 11-27. *Keyword Mapping for the Indicator Section*

Element	Keyword
Display units	indicatorDisplayUnits
Value decimal places	indicatorPrecision
Text size	fontSize
Font color	fontColor
Font family	fontFamily
(Bold)	Bold
(Italic)	Italic
(Underline)	Underline
Horizontal alignment	horizontalAlignment
Vertical alignment	verticalAlignment
Show icon	showIcon
Icon size	iconSize

The *trendline* keyword that describes the trend axis section uses the keywords shown in Table 11-28.

Table 11-28. *Keyword Mapping for the Trend Axis Section*

Element	Keyword
Trend axis	Show
Transparency	Transparency

The *goals* keyword that describes the goals section uses the keywords shown in Table 11-29.

Table 11-29. *Element Keyword Mapping for the Goals Section*

Element	Keyword
Goal	showGoal
Label	goalText
Font color	goalFontColor
Font family	goalFontFamily
Text size	fontSize
(Bold)	goalBold
(Italic)	goalItalic
(Underline)	goalUnderline
Distance	showDistance
Label	distanceLabel
Value decimal places	labelPrecision
(Distance) Font color	distanceFontColor
(Distance) Font family	distanceFontFamily
(Distance) Text size	titleFontSize
Distance direction	Direction

The *status* keyword that describes the color coding section uses the keywords shown in Table 11-30.

Table 11-30. *Keyword Mapping for the Color Coding Section*

Element	Keyword
Direction	Direction
Good color	goodColor
Neutral color	neutralColor
Bad color	badColor

The *distanceLabel* keyword in the JSON for the goal needs one of the attributes shown in Table 11-31.

Table 11-31. *Distance Label Options*

Value	Formatting Popup Option
Value	Value
Percent	Percent
Value, percent	Value, percent

The *direction* keyword in the JSON for the goal needs one of the attributes shown in Table 11-32.

Table 11-32. *Goal Direction Options*

Value	Formatting Popup Option
High is good	Increasing is positive
Low is good	Decreasing is positive

The *direction* keyword in the JSON for the color coding (status) needs one of the attributes shown in Table 11-33.

Table 11-33. *Color Coding Direction Options*

Value	Formatting Popup Option
Positive	High is good
Negative	Low is good

You can find the full JSON code to format a KPI in the sample file Kpi.json.

Paginated Report

The next piece of JSON code that you will see in this chapter is for the paginated report. It is introduced by the *rdlVisual* keyword in a theme file.

Table 11-34 outlines the JSON keywords that you can use when formatting a paginated report and the section that corresponds to each keyword in the Power BI Desktop format pane.

Table 11-34. *Paginated Report Visual Keywords*

Keyword	Formatting Section
general	General
toolbar	Toolbar
autofilter	Export
export	Auto-apply filters
padding	Properties/Padding
title	Title/Title
subTitle	Title/Subtitle
divider	Title/Divider
spacing	Title/Spacing
background	Effects/Background
border	Effects/Visual border

(*continued*)

Table 11-34. (*continued*)

Keyword	Formatting Section
dropShadow	Effects/Shadow
visualHeader	Header Icons
lockAspect	Lock aspect
visualHeaderTooltip	Header Icons/Help tooltip

The code to format a paginated report is as follows:

```
" rdlVisual":
  {"*" :
    {
    "general": [{"responsive": true }],
      "toolbar": [{
        "show": true,
        "paramButton": true,
        "position": 1
        }],
      "autoFilter": [{
        "autoFilter": true
        }],
      "export":  [{
        "exportExcel": false,
        "exportPDF": false,
        "exportAccessiblePDF": false,
        "exportPPTX": false,
        "exportCSV": false,
        "exportWord": false,
        "exportMHTML": false,
        "exportXML": false
        }],
    "title": [{... code not shown ...}],
    "subTitle": [{... code not shown ...}],
    "divider": [{... code not shown ...}],
    "spacing": [{... code not shown ...}],
    "background": [{... code not shown ...}],
    "lockAspect": [{... code not shown ...}],
    "border": [{... code not shown ...}],
    "dropShadow": [{... code not shown ...}],
    "visualHeader": [{... code not shown ...}]
    "visualHeaderTooltip": [{... code not shown ...}]
    "padding": [{... code not shown ...}]
    }
  }
```

The *position* keyword in the JSON for the toolbar needs one of the attributes shown in Table 11-35.

Table 11-35. *Toolbar Position Options*

Value	Formatting Popup Option
0	Top
1	Bottom

Power Apps Visual

Power Apps visuals also let you define only the standard visual elements (background, lock aspect, border, shadow, and the visual header).

The keyword that specifies the formatting attributes of a Python visual is *PowerApps_PBI_CV_C29F1DC C_81F5_4973_94AD_0517D44CC06A.*

Table 11-36 outlines the JSON keywords that you can use when formatting Power Apps visuals and the section that corresponds to each keyword in the Power BI Desktop format pane.

Table 11-36. *Power Apps Visual Keywords*

Keyword	Formatting Section
general	General
padding	Properties/Padding
title	Title/Title
subTitle	Title/Subtitle
divider	Title/Divider
spacing	Title/Spacing
background	Effects/Background
border	Effects/Visual border
dropShadow	Effects/Shadow
visualHeader	Header Icons
lockAspect	Lock aspect
visualHeaderTooltip	Header Icons/Help tooltip

You can find the JSON code for this in the sample file Attributes_PowerApps.json.

The following code snippet provides the JSON code used to pre-format a Power Apps visual. I realize that you have seen it before, but wanted to make the point that common visual elements can be specified for this visual in the theme file even if nothing else can.

```
" PowerApps_PBI_CV_C29F1DCC_81F5_4973_94AD_0517D44CC06A" :
  {"*":
    {
    "general":  [{"responsive": true }],
    "title":  [{... code not shown ...}],
    "subTitle":  [{... code not shown ...}],
    "divider":  [{... code not shown ...}],
    "spacing":  [{... code not shown ...}],
    "background":  [{... code not shown ...}],
    "lockAspect":  [{... code not shown ...}],
    "border":  [{... code not shown ...}],
    "dropShadow":  [{... code not shown ...}],
    "visualHeader":  [{... code not shown ...}]
    "visualHeaderTooltip":  [{... code not shown ...}]
    "padding":  [{... code not shown ...}]
    }
  }
```

Conclusion

In this chapter, you saw how to define formatting in JSON for the gauge, Python visual, R script visual, key influencers visual, decomposition tree visual, KPI, paginated report, and Power Apps visual.

You can now move on to the final set of visual formatting attributes – those used to define actual dashboards themselves and visuals that are not used for data presentation. This is the subject of the following chapter, which is the last of the chapters that covers detailed formatting of visuals.

CHAPTER 12

■ ■ ■

Dashboard Styling

In this chapter, you will finish your tour of the various objects that you can format using theme files. This involves looking at a final couple of visuals and two elements that are not technically visuals at all. These four are

- Action button
- Slicer
- Page
- Filter pane

Sample files that contain the complete JSON code for these elements are available to download from the Apress site.

To conclude the chapter, I also want to describe not only how you can annotate theme files but also how you can actually discover the keywords that Power BI uses for JSON mapping. This will enable you to keep up with changes and developments in Power BI and update your JSON files as the product evolves and new formatting possibilities are added.

Action Button

Action buttons can be formatted just like any other visual in a theme file.

The keyword that specifies the formatting attributes of an action button is *actionButton*.

Table 12-1 outlines the JSON keywords that you can use in action button formatting.

© Adam Aspin 2023

A. Aspin, *Pro Power BI Theme Creation*, https://doi.org/10.1007/978-1-4842-9633-2_12

Table 12-1. *Action Button Keywords*

Keyword	Formatting Section
general	General
shape	Shape
rotation	Rotation
text	Style/Text
icon	Style/Icon
outline	Style/Border
fill	Style/Fill
shadow	Style/Glow
glow	Style/Shadow
visualLink	Action
title	Title
subTitle	Subtitle
divider	Divider
spacing	Spacing
background	Background
lockAspect	Lock aspect
border	Border
dropShadow	Shadow
visualHeader	Header icons
visualHeaderTooltip	Tooltips

The following code snippet gives the JSON theme for action button formatting. It is surprisingly long, as several elements can be formatted differently according to the button state. This effectively multiplies certain subsections of the code.

```
"actionButton":
  {"*":
    {
      "shape": [{
              "$id": "default",
              "tileShape": "heart",
              "roundEdge": 5,
              "angle": 45,
              "arrowheadSize": 4,
              "hexagonSlant": 40,
              "octagonSnipSize": 40,
```

```
            "tabCutCornerSnipSizeTopRight": 40,
            "tabCutCornerSnipSizeTopRight": 40
        }],
"rotation": [{
        "angle": 45,
        "shapeAngle": 50,
        "textAngle": 35
        }],
"text": [{
        "show": true}
        ,
        {
        "$id": "default",
        "text": "Button Text",
        "fontColor": {"solid": {"color": "#808080"}},
        "padding": 3,
        "verticalAlignment": "middle",
        "horizontalAlignment": "center",
        "fontSize": 10,
        "fontFamily": "Arial",
        "bold": true,
        "italic": false,
        "underline": false,
        "topMargin": 3,
        "bottomMargin": 3,
        "leftMargin": 3,
        "rightMargin": 3
        }
        ,
        {
        "$id": "hover",
        "text": "Hover button Text",
        "fontColor": {"solid": {"color": "#808080"}},
        "padding": 3,
        "verticalAlignment": "middle",
        "horizontalAlignment": "center",
        "fontSize": 10,
        "fontFamily": "Arial",
        "bold": true,
        "italic": false,
        "underline": false,
        "topMargin": 3,
        "bottomMargin": 3,
        "leftMargin": 3,
        "rightMargin": 3
        }
        ,
        {
        "$id": "selected",
        "text": "Press button Text",
        "fontColor": {"solid": {"color": "#808080"}},
```

```
        "padding": 3,
        "verticalAlignment": "middle",
        "horizontalAlignment": "center",
        "fontSize": 10,
        "fontFamily": "Arial",
        "bold": true,
        "italic": false,
        "underline": false,
        "topMargin": 3,
        "bottomMargin": 3,
        "leftMargin": 3,
        "rightMargin": 3
      }
      ,
      {
      "$id": "disabled",
      "text": "Disabled button Text",
      "fontColor": {"solid": {"color": "#808080"}},
      "padding": 3,
      "verticalAlignment": "middle",
      "horizontalAlignment": "center",
      "fontSize": 10,
      "fontFamily": "Arial",
      "topMargin": 3,
      "bottomMargin": 3,
      "leftMargin": 3,
      "rightMargin": 3
      }],
  "icon": [{
      "show": true
      },
      {
      "$id": "default",
      "shapeType": "help",
      "padding": 3,
      "verticalAlignment": "middle",
      "horizontalAlignment": "center",
      "lineColor": {"solid": {"color": "#F5F4F0"}},
      "lineTransparency": 10,
      "lineWeight": 2,
      "topMargin": 3,
      "bottomMargin": 3,
      "leftMargin": 3,
      "rightMargin": 3,
      "placement": "left",
      "iconSize": 22
      },
      {
      "$id": "hover",
      "shapeType": "help",
      "padding": 3,
```

```
        "verticalAlignment": "middle",
        "horizontalAlignment": "center",
        "lineColor": {"solid": {"color": "#F5F4F0"}},
        "lineTransparency": 10,
        "lineWeight": 2,
        "topMargin": 3,
        "bottomMargin": 3,
        "leftMargin": 3,
        "rightMargin": 3,
        "placement": "left",
        "iconSize": 22
        },
        {
        "$id": "selected",
        "shapeType": "help",
        "padding": 3,
        "verticalAlignment": "middle",
        "horizontalAlignment": "center",
        "lineColor": {"solid": {"color": "#F5F4F0"}},
        "lineTransparency": 10,
        "lineWeight": 2,
        "topMargin": 3,
        "bottomMargin": 3,
        "leftMargin": 3,
        "rightMargin": 3,
        "placement": "left",
        "iconSize": 22
        },
        {
        "$id": "disabled",
        "shapeType": "help",
        "padding": 3,
        "verticalAlignment": "middle",
        "horizontalAlignment": "center",
        "lineColor": {"solid": {"color": "#F5F4F0"}},
        "lineTransparency": 10,
        "lineWeight": 2,
        "topMargin": 3,
        "bottomMargin": 3,
        "leftMargin": 3,
        "rightMargin": 3,
        "placement": "left",
        "iconSize": 22
        }],
    "outline": [{
        "show": true
        },
        {
        "$id": "default",
        "lineColor": {"solid": {"color": "#F5F4F0"}},
        "transparency": 10,
```

```
                    "weight": 2
                },
                {
                "$id": "hover",
                "lineColor": {"solid": {"color": "#F5F4F0"}},
                "transparency": 5,
                "weight": 2
                },                    {
                "$id": "selected",
                "lineColor": {"solid": {"color": "#F5F4F0"}},
                "transparency": 15,
                "weight": 2
                },                    {
                "$id": "disabled",
                "lineColor": {"solid": {"color": "#F5F4F0"}},
                "transparency": 20,
                "weight": 2
                }],
        "fill":
                [{
                "show":true
                },
                {
                "$id": "default",
                "transparency": 0,
                "fillColor": {"solid":{"color":"#949494"}},
                "scaling": "Fill"
                },
                {
                "$id": "hover",
                "transparency": 0,
                "fillColor": {"solid":{"color":"#949494"}},
                "scaling": "Fill"
                },
                {
                "$id": "selected",
                "transparency": 0,
                "fillColor": {"solid":{"color":"#111111"}},
                "scaling": "Fill"
                },
                {
                "$id": "disabled",
                "transparency": 0,
                "fillColor": {"solid":{"color":"#111111"}},
                "scaling": "Fill"
                }],
        "shadow":
                [{
                "show":true
                },
```

```
{
"$id": "default",
"transparency": 0,
"color": {"solid":{"color":"#949494"}},
"shadowBlur": 25,
"shadowPositionPreset": "bottom"
},
{
"$id": "hover",
"transparency": 0,
"color": {"solid":{"color":"#949494"}},
"shadowBlur": 25,
"shadowPositionPreset": "bottom"
},
{
"$id": "selected",
"transparency": 0,
"color": {"solid":{"color":"#111111"}},
"shadowBlur": 25,
"shadowPositionPreset": "bottom"
},
{
"$id": "disabled",
"transparency": 0,
"color": {"solid":{"color":"#111111"}},
"shadowBlur": 25,
"shadowPositionPreset": "bottom"
}],
"glow":
[{
"show":true
},
{
"$id": "default",
"transparency": 0,
"color": {"solid":{"color":"#949494"}},
"shadowBlur": 25
},
{
"$id": "hover",
"transparency": 0,
"color": {"solid":{"color":"#949494"}},
"shadowBlur": 25
},
{
"$id": "selected",
"transparency": 0,
"color": {"solid":{"color":"#111111"}},
"shadowBlur": 25
},
```

239

```
            {
            "$id": "disabled",
            "transparency": 0,
            "color": {"solid":{"color":"#111111"}},
            "shadowBlur": 25
            }],
      "visualLink": [{
            "show": true,
            "type": "Bookmark",
            "tooltip": "Tooltip text here"
            }],
      "title": [{... code not shown ...}],
      "subTitle": [{... code not shown ...}],
      "divider": [{... code not shown ...}],
      "spacing": [{... code not shown ...}],
      "background": [{... code not shown ...}],
      "lockAspect": [{... code not shown ...}],
      "border": [{... code not shown ...}],
      "dropShadow": [{... code not shown ...}],
      "visualHeader": [{... code not shown ...}]
      "visualHeaderTooltip": [{... code not shown ...}]
      "padding": [{... code not shown ...}]
      }
   }
```

It is worth noting to begin with that you can set the attributes for icon, fill, and text to reflect different settings for the four following button states:

- Default

- Hover

- Press

- Disabled

The fact that a new concept is introduced – button state – has an impact on the object nesting in the theme file JSON. In effect, an added level of nesting is applied inside each main element of the action button.

Each of these states is defined using the *$id* keyword in the JSON that defines the button state. For the two sections in the JSON that specify an *$id* keyword (*text* and *icon*), you need one of the values shown in Table 12-2. The outcome of this is that you have *four* nearly identical subsections in the JSON that describe text and icon formatting – one for each button state.

Table 12-2. *Button State Options*

Value	Formatting Popup Option
default	Default state
hover	On hover
selected	On press
disabled	Disabled

The *text* keyword that describes the button text section uses the keywords shown in Table 12-3.

Table 12-3. *Keyword Mapping for the Button Text Section*

Element	Keyword
Button text	Show
Button text	Text
Font color	fontColor
Padding	Padding
Vertical alignment	verticalAlignment
Horizontal alignment	horizontalAlignment
Text size	fontSize
Font family	fontFamily

The *icon* keyword that describes the icon section uses the keywords shown in Table 12-4.

Table 12-4. *Keyword Mapping for the Icon Section*

Element	Keyword
Icon	Show
Shape	shapeType
Padding	Padding
Vertical alignment	verticalAlignment
Horizontal alignment	horizontalAlignment
Line color	lineColor
Transparency	lineTransparency
Weight	lineWeight

The *outline* keyword that describes the outline section uses the keywords shown in Table 12-5.

Table 12-5. *Keyword Mapping for the Outline Section*

Element	Keyword
Outline	Show
Outline color	lineColor
Transparency	Transparency
Outline weight	Weight
Round edges	roundEdge

The *fill* keyword that describes the fill section uses the keywords shown in Table 12-6.

Table 12-6. *Keyword Mapping for the Fill Section*

Element	Keyword
Fill	Show
Fill color	fillColor
Transparency	Transparency

The *shadow* keyword that describes the Shadow section uses the keywords shown in Table 12-7.

Table 12-7. *Keyword Mapping for the Shadow Section*

Element	Keyword
Shadow	Show
Transparency	Transparency
Color	Color
Blur	shadowBlur
Position	shadowPositionPreset

The *glow* keyword that describes the Glow section uses the keywords shown in Table 12-8.

Table 12-8. *Keyword Mapping for the Glow Section*

Element	Keyword
Glow	Show
Transparency	Transparency
Color	Color
Blur	shadowBlur

The *visualLink* keyword that describes the button Action section uses the keywords shown in Table 12-9.

Table 12-9. *Keyword Mapping for the Button Action Section*

Element	Keyword
Action	Show
Type	Type
Text	Tooltip

Table 12-10 shows the vertical alignment attributes for icons.

Table 12-10. *Vertical Alignment Options*

Value	Formatting Popup Option
top	Top
middle	Middle
bottom	Bottom

Table 12-11 shows the horizontal alignment attributes for icons.

Table 12-11. *Horizontal Alignment Options*

Value	Formatting Popup Option
left	Left
center	Center
right	Right

Shadow position options are given in Table 5-14.
You can find the JSON code for action buttons in the sample file Attributes_ActionButton.json.

Slicer

As you might expect, you can also format slicers in a theme file.
The keyword that specifies the formatting attributes of a slicer is *slicer*.
Table 12-12 outlines the JSON keywords that you can use in slicer formatting.

Table 12-12. *Slicer Keywords*

Keyword	Formatting Section
General	General
slicerHeader	Slicer header
numericInputStyle	Numeric inputs
Slider	Slider
Selection	Selection controls
Header	Header
Date	Date
Items	Items
Title	Title
subtitle	Subtitle

(continued)

Table 12-12. (*continued*)

Keyword	Formatting Section
Divider	Divider
Spacing	Spacing
Background	Background
lockAspect	Lock aspect
Border	Border
dropShadow	Shadow
visualHeader	Header icons
visualHeaderTooltip	Tooltips

You need to remember that not all slicer section definition elements will be visible at once in Power BI Desktop, as they depend on the type of slicer that you have defined. You can, nonetheless, specify all the available options in the theme file using the appropriate keyword.

The following code snippet shows the JSON used to format a slicer:

```
"slicer":
  {"*":
    {
      "data": [{
            "mode": "Before"
            }],
      "pendingChangesIcon": [{
            "show": true,
            "position": "left",
            "size": 12,
            "color": {"solid": {"color": "#F5F4F0"}},
            "transparency": 0,
            "tooltipText": "Awaiting final confirmation"
            }],
      "selection": [{
            "show": true,
            "singleSelect": true,
            "strictSingleSelect": false,
            "selectAllCheckboxEnabled": false
            }],
      "header":    [{
            "show": false,
            "text": "Slicer Text",
            "fontColor": {"solid": {"color": "#ffffff"}},
            "background": {"solid": {"color": "#7A8C97"}},
            "outline": "None",
            "textSize": 9,
            "fontFamily": "Arial",
            "bold": false,
            "italic": false,
```

```
            "underline": false
            }],
    "numericInputStyle":[{
            "fontColor": {"solid": {"color": "#ffffff"}},
            "backgroundColor": {"solid": {"color": "#B7B7B7"}},
            "fontSize": 9,
            "fontFamily": "Arial"
            }],
    "slider":    [{
            "color": {"solid": { "color": "#F5F4F0"} }
            }],
    "date":[{
            "fontColor": {"solid": {"color": "#6e6e6e"}},
            "background": {"solid": {"color": "#cfcfcf"}},
            "fontFamily": "Arial",
            "textSize": 8
            }],
    "items":    [{
            "fontColor": {"solid": {"color": "#F5F4F0"}},
            "backgroundColor": {"solid": {"color": "#B7B7B7"}},
            "outline": "None",
            "textSize": 8,
            "fontFamily": "Arial",
            "bold": false,
            "italic": false,
            "underline": false
            }],
    "title": [{... code not shown ...}],
    "subTitle": [{... code not shown ...}],
    "divider": [{... code not shown ...}],
    "spacing": [{... code not shown ...}],
    "background": [{... code not shown ...}],
    "lockAspect": [{... code not shown ...}],
    "border": [{... code not shown ...}],
    "dropShadow": [{... code not shown ...}],
    "visualHeader": [{... code not shown ...}]
    "visualHeaderTooltip": [{... code not shown ...}]
    "padding": [{... code not shown ...}]
    }
}
```

The *general* keyword that describes the selection controls section uses the keywords shown in Table 12-13.

Table 12-13. *Keyword Mapping for the General Section*

Element	Keyword
Outline color	outlineColor
Outline weight	outlineWeight
Orientation	Orientation
Responsive	Responsive

The *selection* keyword that describes the selection controls section uses the keywords shown in Table 12-14.

Table 12-14. *Keyword Mapping for the Selection Controls Section*

Element	Keyword
Selection controls	Show
Single select	singleSelect
Multi-select with CTRL	strictSingleSelect
Show "Select all" option	selectAllCheckboxEnabled

The *header* keyword that describes the header section uses the keywords shown in Table 12-15.

Table 12-15. *Keyword Mapping for the Slicer Header Section*

Element	Keyword
Slicer header	show
Title text	text
Font color	fontColor
Background	background
Outline	outline
Text size	textSize
Font family	fontFamily
(Bold)	bold
(Italic)	italic
(Underline)	underline

The *numericInputStyle* keyword that describes the numeric inputs section uses the keywords shown in Table 12-16.

Table 12-16. *Keyword Mapping for the Numeric Inputs Section*

Element	Keyword
Font color	fontColor
Background	backgroundColor
Text size	textSize
Font family	fontFamily

The *slider* keyword that describes the slider section uses the keywords shown in Table 12-17.

Table 12-17. *Element Keyword Mapping for the Slider Section*

Element	Keyword
Slider	Show
Color	Color

The *items* keyword that describes the items section uses the keywords shown in Table 12-18.

Table 12-18. *Keyword Mapping for the Items Section*

Element	Keyword
Font color	fontColor
Background	backgroundColor
Outline	Outline
Text size	textSize
Font family	fontFamily

The *pendingChangesIcon* keyword that describes the Pending icon section uses the keywords shown in Table 12-19.

Table 12-19. *Keyword Mapping for the Pending Changes Section*

Element	Keyword
Pending icon	Show
Position	Position
Size	size
Color	color
Transparency	transparency
Tooltip text	tooltipText

Table 12-20 shows the position attributes for the Horizontal Alignment icon.

Table 12-20. *Horizontal Alignment Options*

Value	Formatting Popup Option
Left	left
Right	right

You can find the JSON code for this in the sample file Attributes_Slicer.json.
The header and items outline options are given in Table 6-10.
The orientation options are given in Table 7-15.

Filter Pane

It may seem surprising – after all, it is not a visual – but you can also format the filter pane in a theme file just as you can format it interactively in Power BI Desktop.

There is no single keyword that introduces the formatting attributes of a filter pane. This is because the filter pane formatting is set at the level of generic formatting. You first met this concept in Chapter 4. Indeed, it could be a good idea to refer back to Chapter 4 for a quick refresher on the concept of generic formatting should you need to.

Filter pane formatting is defined using two keywords at this level. They are

- outspacePane

- filterCard

Table 12-21 outlines the JSON keywords that you can use in filter pane formatting.

Table 12-21. *Filter Pane Keywords*

Keyword	Formatting Section
outspacePane	Filter pane
filterCard	Filter card

```
"visualStyles":
  {"*":
    {"*":
      {
      "outspacePane": [{
      "backgroundColor": {"solid": {"color": "#999999"}},
      "transparency": 30,
      "foregroundColor": {"solid": {"color": "#374649"}},
      "titleSize": 18,
      "headerSize": 12,
      "fontFamily": "Arial",
      "border": true,
      "borderColor": {"solid": {"color": "#909090"}},
      "width": 190,
      "checkboxAndApplyColor": {"solid": {"color": "#374649"}},
      "searchTextSize": 11,
      "inputBoxColor": {"solid": {"color": "#ffffff"}}
      }],
    "filterCard": [{
      "$id": "Applied",
      "backgroundColor": {"solid": {"color": "#171796"}},
      "transparency": 0,
      "border": true,
      "borderColor": {"solid": {"color": "#909090"}},
      "foregroundColor": {"solid": {"color": "#ffffff"}},
      "textSize": 10,
      "fontFamily": "Arial",
      "inputBoxColor": {"solid": {"color": "#C8C8C8"}}
      },
```

```
        {
        "$id": "Available",
        "backgroundColor": {"solid": {"color": "#171796"}},
        "transparency": 40,
        "border": true,
        "borderColor": {"solid": {"color": "#909090"}},
        "foregroundColor": {"solid": {"color": "#ffffff"}},
        "textSize": 9,
        "fontFamily": "Arial",
        "inputBoxColor": {"solid": {"color": "#777777"}}
        }]
          }
        }
      }
    }
  }
```

The *outspacePane* keyword that describes the filter pane section uses the keywords shown in Table 12-22.

Table 12-22. *Keyword Mapping for the Filter Pane Section*

Element	Keyword
Background color	backgroundColor
Transparency	transparency
Font and icon color	foregroundColor
Title text size	titleSize
Header text size	headerSize
Font family	fontFamily
Border	border
Border color	borderColor
Width	width
Checkbox and apply color	checkboxAndApplyColor
Search size text	searchTextSize
Input box color	inputBoxColor

Before looking at these elements and keywords, it is worth noting that there are two types of filter card:

- Applied
- Not applied

These represent, respectively, cards where filters have or have not been applied.

The two filter card states are introduced using the $id keyword. This means that JSON code is duplicated for each filter state– as was the case for action button states, as you saw earlier in this chapter.

The *$id* keyword in the JSON that defines the type of a filter card uses the keywords shown in Table 12-23.

Table 12-23. *Button State Options*

Value	Formatting Popup Option
Applied	Applied
Available	Available

The *filterCard* keyword that describes the filter card section uses the keywords shown in Table 12-24.

Table 12-24. *Keyword Mapping for the Filter Card Section*

Element	Keyword
Background color	backgroundColor
Transparency	Transparency
Border	Border
Border color	borderColor
Font and icon color	foregroundColor
Text size	textSize
Font family	fontFamily
Input box color	inputBoxColor

You can find the JSON code for this in the sample file Attributes_FilterCard.json.

Page Formatting

The page background can also be formatted using a theme file. Rather like the format pane, this is considered (as far as the theme file goes) as a generic rather than a specific formatting element and so appears as part of the generic visual style in the file.

Table 12-25 outlines the JSON keywords that you can use for page formatting and the section that corresponds to each keyword in the Power BI Desktop format pane.

Table 12-25. *Page Formatting Keywords*

Keyword	Formatting Section
background	Page background
displayArea	Page alignment
outspace	Wallpaper

The code for page formatting is found in the sample file Attributes_Page.json. You can see the complete code (to explain the position of page formatting in the file structure) in the following code snippet:

```
{
"name": "Page formatting",
"dataColors": ["#12239E", "#118DFF", "#00FF29", "#6B007B", "#FFB000", "#F5F4F0", "#FFFF00",
"#FF001E"],
"visualStyles":
  {
  "*":
    {"*":
      {
     "page":
       {"*":
         {
         "background":
           [{
           "color": {"solid": {"color": "#F5F4F0"}},
           "transparency": 0
           }],
         "displayArea":
           [{
           "verticalAlignment": "Middle"
           }],
         "outspace":
           [{
           "color": {"solid": {"color": "#F5F4F0"}}
           "transparency": 20
           }]
         }
       }
     }
   }
  }
}
```

The *background* keyword that describes the page background section uses the keywords shown in Table 12-26.

Table 12-26. *Keyword Mapping for Page Background*

Element	Keyword
Color	Color
Transparency	Transparency

The *displayArea* keyword that describes the page alignment section uses the keyword shown in Table 12-27.

Table 12-27. *Keyword Mapping for Page Alignment*

Element	Keyword
Vertical alignment	verticalAlignment

251

The *outspace* keyword that describes the *wallpaper* section uses the keywords shown in Table 12-28.

Table 12-28. *Keyword Mapping for Wallpaper*

Element	Keyword
Color	Color
Transparency	Transparency

The *verticalAlignment* keyword in the JSON for setting the page alignment needs one of the attributes shown in Table 12-29.

Table 12-29. *Legend Position Options*

Value	Formatting Popup Option
Top	Top
Middle	Middle

Adding Notes to Theme Files

Unfortunately, the JSON used in theme files does not support adding comments as you can with most programming languages. So you cannot use // or /* ...*/ or - - to introduce comments in or after JSON code as you can in certain programming languages.

While this can certainly be a limitation, there is a workaround that you might find useful. This is to invent your own keywords and use them to introduce your comments to annotate the file. You can see an example of this in the following code snippet, where a couple of notes are added at the top of the visual styles formatting section of the theme file:

```
"visualStyles":
{
"myNotesForOneSubject" : "Currently you can add notes like this",
"Notes" : "It apparently has to be here in the structure",
```

Another trick is to add notes at the start of the formatting for a specific visual – like this:

```
"textbox":
  {"*" :
    {
      "myTextboxNotes" : "Currently you can add notes like this",
      "title":    [{ ... CODE OMITTED ... }]
```

You need to be aware that this is unsupported, and although it works for the moment, it might cease to work in the future. Also, I advise you to invent keywords that are highly unlikely to be used one day by Power BI. This way, you minimize the risk of causing failures as the product evolves.

Another idea that can be useful when you create multiple theme files is to use the name field to add header information to a theme file. You can see an example of this in the following code snippet:

```
"name": "Generic Base Formatting add lots of comments to explain why this is used, when it
was created etc"
```

However, you cannot add line or carriage returns inside the comment itself. And I prefer to avoid single or double quotes as well. Nonetheless, adding comments at the top of a theme file can be particularly useful if your collection of theme files starts to grow, as it enables you to track the files much more easily than through reading the JSON when trying to remember what the specifics of a theme are.

Analyze the Contents of Power BI Desktop Files

As Power BI evolves, new features will, inevitably, be added, and this means additional formatting possibilities. So you might need to discover new keywords and/or their attributes.

The way to do this is to

1. Create a Power BI Desktop file containing the visual whose JSON formatting you want to extend.

2. Manually format the attributes that are new (this means that they will appear in the Power BI Desktop file – default formatting does not).

3. Save the Power BI Desktop file.

4. Unzip the Power BI Desktop file to a folder.

5. Open the folder where you unzipped the Power BI Desktop file.

6. Open the folder named Report.

7. Open the file named Layout with a text editor.

You will see a – largely incomprehensible – text that, in fact, contains the JSON formatting for the elements that you formatted in step 2. The trick now is to find what you formatted and consequently the keyword that Power BI uses for the theme file. You need to be warned that this can take practice and patience!

You can see an example of the report contents in Figure 12-1. Note that the outspacePane keyword is highlighted so you can see how its properties appear in the underlying Power BI Desktop file.

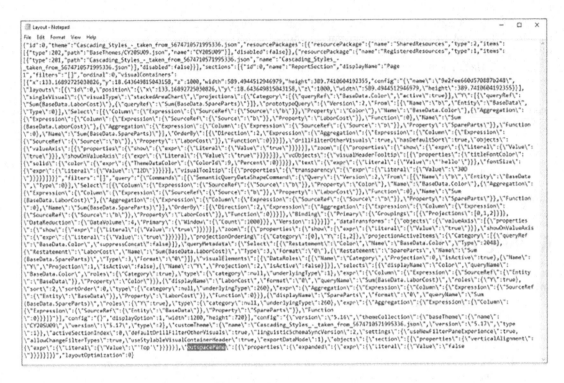

Figure 12-1. *Partial contents of a .PBIX file*

Conclusion

This concludes your overview of the various objects that can be formatted using a JSON theme file in Power BI. In this chapter, you saw how to format action buttons, slicers, the page itself, and the filter pane using the same approaches that you have applied in previous chapters.

I have now only one final process to explain – how to create cascading styles in theme files. You will learn about this in the next – and final – chapter.

CHAPTER 13

■ ■ ■

Cascading Styles

In this book, you have seen that you can define visual appearance using JSON styles at several levels. The final part of the learning curve is to appreciate how styles can cascade from a high-level definition down through progressively finer specifications. Appreciating this will allow you to save time when creating theme files as you are, in effect, able to define many presentation attributes once – rather than have to define all the subtleties of every visual.

These "cascading styles" happen in two separate ways:

- Text definitions cascade down through generic style definitions to specific visual formatting definitions.

- Colors cascade down from the color palette settings to be used wherever a color can be applied in a theme file.

Style Hierarchy

The first thing to remember is that you can define presentation elements at four possible levels. These are

- Element level

- Text class

- Generic visual style

- Specific visual style

There is a strict hierarchy in this approach. That is, any specification set at the element level will remain valid until overwritten by a text class. Text classes can be further refined by defining settings for a generic visual style which, in turn, can be redefined as part of a specific visual style.

This cascading style approach is possible because the style hierarchy progresses from the wide-ranging to the highly restrictive. If you remember from Chapter 2, defining a first-level element covers a range of visual attributes. In Chapter 3, you saw that text classes are more focused – and this continues right down to the detailed definition of one aspect of a specific visual.

However, you need to be aware that the hierarchy is not always complete. By this I mean that not everything can be set at the element level (there are many aspects of visual formatting that are not covered by the element options). Equally, text classes do not cover all aspects of formatting – and there is some overlap between these two levels.

As with all facets of theme file creation, this is best appreciated with an example.

Take, for instance, the totals that you can format in a table. These can be defined (in whole or in part) at three of the four levels described earlier. Table 13-1 describes this.

© Adam Aspin 2023
A. Aspin, *Pro Power BI Theme Creation*, https://doi.org/10.1007/978-1-4842-9633-2_13

Table 13-1. *Column Header Formatting*

Level	Definition
Text class	Label
Generic visual style	Total
Specific visual style	tableEx

The *textClasses* section lets you specify a *label* element. This covers (among other things) totals in tables and matrices. If you specify a font size and color here, it will set (among other things) the color of the totals used in tables and matrices.

You can then choose to set a few more specific elements (such as the font color and size) at the level of generic definitions in a theme file. This is done using the *total* keyword and the relevant elements. This will set totals wherever this keyword is used in visuals, which means the table and matrix visuals, and will override anything (including, in this example, the font size and color) set using text classes. It will *not*, however, override any other elements that are set using the *label* text class.

Finally, you can set formatting that applies only to table visuals. You saw this in Chapter 6. Any specifics here will override the generic formatting for totals – but *only in tables, not in matrices*.

Figure 13-1 shows this more visually.

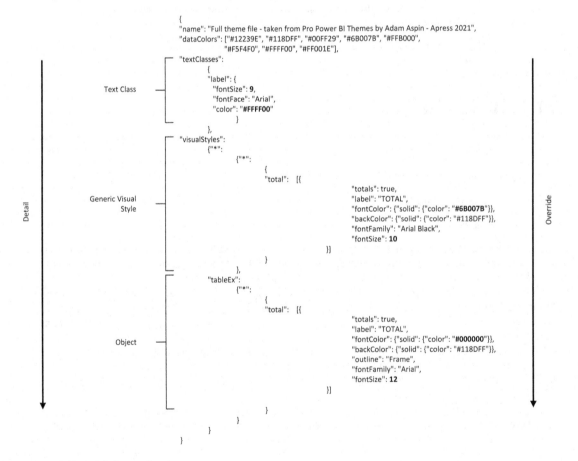

Figure 13-1. *Style hierarchy*

You can see in Figure 13-1 that the colors are set three times:

- FFFF00 (yellow) at the text class level
- F5F4F0 (orange) at the generic level for totals of tables and matrices
- 6B007B (purple) for table totals

You will also note that the font size is set three times:

- 9 point at the text class level
- 10 point at the generic level for totals of tables and matrices
- 12 point (black) for table totals

Totals for tables and matrices are set by the *label* text class. This also sets the font face and size for totals in tables and matrices as well as for other elements (the full list is also available in Chapter 2).

Totals for tables and matrices are then *overridden* – as well as having other aspects of their presentation defined – as part of the generic style definition (remember this is nested inside an asterisk object and not a specific visual object).

Finally, totals for tables (but not matrices) are defined as you saw explained in Chapter 6. This takes place at the deepest, most detailed level.

You may well be asking what this means in practice. So let's take a look at the effects produced by applying progressively more detailed formatting that you can see in the JSON theme file CascadingText.json.

To begin with, here is the JSON that only applies a text class – *label* in this instance. This class sets the formatting for the totals in tables and matrices, as you can see in Figure 13-2, where the totals are yellow and 9 point.

```
{
"name": "Cascading Styles - taken from Pro Power BI Themes by Adam Aspin - Apress 2021",
"dataColors": ["#12239E", "#118DFF", "#00FF29", "#6B007B", "#FFB000", "#F5F4F0", "#FFFF00", "#FF001E"],
"textClasses":
        {
        "label": {
          "fontSize": 9,
          "fontFace": "Arial",
          "color": "#FFFF00"
                }
        }
}
```

CountryName	CostPrice	LaborCost	SpareParts
France	1460100	36471	42430
Germany	160000	1461	1600
Spain	178700	2272	4420
Switzerland	701150	27207	30315
United Kingdom	10193880	129952	227250
USA	7716065	113135	189045

Figure 13-2. *Applying a text class*

In Figure 13-3, the JSON is augmented by adding a generic style for *total*. This has the effect of overriding the font attributes that are set by the text class. You can see this in Figure 13-4 where the totals are orange and 10 point.

```
{
"name": "Cascading Styles - taken from Pro Power BI Themes by Adam Aspin - Apress 2021",
"dataColors": ["#12239E", "#118DFF", "#00FF29", "#6B007B", "#FFB000", "#F5F4F0", "#FFFF00", "#FF001E"],
"textClasses":
        {
        "label": {
          "fontSize": 9,
          "fontFace": "Arial",
          "color": "#FFFF00"
               }
        },
"visualStyles":
        {"*":
            {"*":
                {
                "total":  [{

                              "totals": true,
                              "label": "TOTAL",
                              "fontColor": {"solid": {"color": "#FFB000"}},
                              "fontFamily": "Arial",
                              "fontSize": 10

                          }]

                }
            }
        }
}
```

Color	CostPrice	DeliveryCharge	LaborCost
Black	2096930	22275	31563
Blue	2417900	25375	35615
British Racing Green	2068590	23225	24959
Canary Yellow	3217945	40775	51983
Dark Purple	1606540	16500	25305
Green	1503145	18590	29921
Night Blue	1451370	13675	20121
Red	3618405	47750	55281
Silver	2429070	31805	35750
TOTAL	20409895	239970	310498

Figure 13-3. Applying a generic style

It is worth noting, however, that any other attributes set by the text class are not affected by the generic style definition.

Finally, we can apply the full JSON that was shown previously in Figure 13-1. This overrides the formatting of totals – but only for *tables*, and *not for matrices*. You can see this in Figure 13-4 where the totals are purple and 12 point for the *table only*.

Table				Matrix			
Color	CostPrice	DeliveryCharge	LaborCost	Color	CostPrice	DeliveryCharge	LaborCost
Black	2096930	22275	31563	Black	2096930	22275	31563
Blue	2417900	25375	35615	Blue	2417900	25375	35615
British Racing Green	2068590	23225	24959	British Racing Green	2068590	23225	24959
Canary Yellow	3217945	40775	51983	Canary Yellow	3217945	40775	51983
Dark Purple	1606540	16500	25305	Dark Purple	1606540	16500	25305
Green	1503145	18590	29921	Green	1503145	18590	29921
Night Blue	1451370	13675	20121	Night Blue	1451370	13675	20121
Red	3618405	47750	55281	Red	3618405	47750	55281
Silver	2429070	31805	35750	Silver	2429070	31805	35750
TOTAL	20409895	239970	310498	Total	20409895	239970	310498

Figure 13-4. *Applying a specific style*

Defining a structured theme file that mixes formatting at multiple levels can be a considerable investment in time and effort. However, the up-front investment certainly makes theme updates much easier and considerably faster. This is because you can often reset a swathe of attributes for multiple visuals with a single change at the level of generic formatting – while leaving highly specific formatting of certain visuals as is. This can also ensure a high level of presentational conformity in your dashboards.

Should you decide to adopt this way of creating theme files, I have a few suggestions that you may find useful in practice:

- Concentrate on the generic level of theme definition – and even disregard text classes altogether. This approach lets you define standardized attributes in a clear way that is easy to understand and review.

- Only apply attributes at one level unless you are deliberately adding an override at a lower level for a specific visual. In other words, avoid duplication.

- Only apply the elements that you really need to define globally at the generic level. There is no point in setting everything you can and then overriding most of it at the level of specific visuals.

- When defining generic formatting, there is often little point in setting every attribute for an element. Here again, only apply the attributes that you need to set globally.

Putting It All Together

Theme creation is essentially a practical matter. So a concrete example of a couple of theme files should help you better understand the difference between a "standard" theme file (where everything is defined individually) and a cascading theme file.

If you look in the downloadable JSON files, you will find

- FullTheme.json: This file contains the complete formatting definition for all visuals. Only page and filter pane definitions are set at the generic level.

- CascadingTheme.json: This is an example of how you can define a series of generic elements followed by highly specific custom formatting for certain visuals.

- FullThemeWithClasses.json: This file contains the complete formatting definition for all visuals. Only page and filter pane definitions are set at the generic level. However, colors and text classes are also added to the file.

I am not going to display the JSON for all three here as together they contain several thousand lines of JSON. However, I do suggest that you take a look in order to understand the different approaches. They may well also be useful to you as a basis for adapting your own theme files.

Applying Cascading Palette Colors

So far in this book, whenever colors have been applied, you have seen a "hard-coded" hexadecimal value. While this is certainly the easiest way to set colors, doing this can make it laborious to change colors in an entire theme file. This is particularly true if you have created a complete set of JSON that specifies every possible object. Indeed, this approach can lead to hours of copy/replace work to reset colors. This is usually followed by hours of testing dozens of visual types once the themes have been applied.

Fortunately, colors can also cascade through a theme file. This makes modifying color settings much easier in practice.

Setting up cascading color settings works like this:

- *First*, set the color palette by defining the eight variable colors as you would normally. This means that you have a choice of 60 colors to choose from (ten columns by six rows in the color palette). Not only do you have a range of colors that you can use, but they are variations on ten core colors – which means that color choices are usually more harmonious.

- *Second*, refer to these colors *not by their hexadecimal references* but by their *position in the palette*.

It is the second part that is new, and here is how it works.
Wherever you would normally use a color reference like this:

```
{"solid": {"color": "#999999"}}
```

To ensure cascading color references, you would use JSON like this instead:

```
{"solid": {"color": {"expr":{"ThemeDataColor":{"ColorId": 9, "Percent": -0.5}}}}}
```

As this piece of JSON is somewhat unintelligible at first sight, take a look at its structure.
The hexadecimal reference is the only part that changes – it becomes

```
{"expr":{"ThemeDataColor":{"ColorId": 9, "Percent": -0.5}}}
```

The three keywords used here are explained in Table 13-2.

Table 13-2. *Palette Color Cascade References*

Keyword	Definition
ThemeDataColor	Indicates that the palette will be used
ColorId	Defines the column to use in the palette
Percent	Applies a percentage intensity to the selected color

The nested structure looks like that shown as follows:

```
{"solid":
    {"color":
      {"expr":
       {"ThemeDataColor":
          {"ColorId": 5, "Percent": -0.25}
       }
      }
    }
}
```

The two main things that need a little explanation are

- ColorId
- Percent

The *ColorId* is a numeric value that represents the column in the palette. It is zero based and applies from left to right. This means that the columns are numbered 0 to 9 for all 10 columns of colors.

So, interestingly, the JSON keyword ThemeDataColor 0 corresponds to the first column in the color palette (defined as white), ThemeDataColor 1 corresponds to the second column in the color palette (defined as black), ThemeDataColor 2 corresponds to the third column in the color palette (which the palette popup, confusingly, describes as Theme color 1), and so on. This is something that you end up getting used to.

Equally, in the color palette, the *shades* of each palette color are defined as (from top to bottom)

- 60% lighter
- 40% lighter
- 20% lighter
- 25% darker
- 50% darker

These are represented by the figure for the JSON *Percent* keyword. However (and also somewhat confusingly), the lighter shades of the palette are set as positive figures, and the darker shades are set as negative figures.

This mapping is described in Figure 13-5.

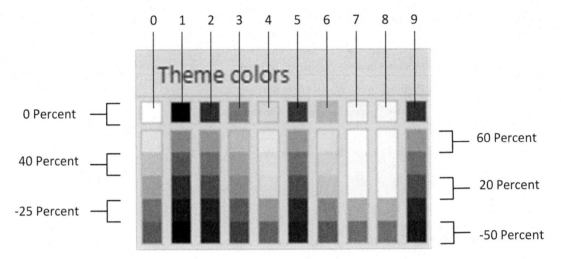

Figure 13-5. *Palette correspondence to theme data colors*

Selecting a color from one of the rows in the color palette requires you to set a percentage figure corresponding to the required shade. As you can see, the lighter shades are set as positive values, and the darker shades as negative values.

As an example, to apply the second row of the color palette in the penultimate column, the percentage figure will be 0.6, as in the following code snippet:

```
{"solid": {"color": {"expr":{"ThemeDataColor":{"ColorId": 8, "Percent": 0.6}}}}}
```

To apply the bottom row of the seventh column (and the fifth that you can set), the percentage figure will be -0.5, as in the following code snippet:

```
{"solid": {"color": {"expr":{"ThemeDataColor":{"ColorId": 6, "Percent": -0.5}}}}}
```

To stress the point, the lighter shades are positive figures, and darker shades are negative figures.

If you want to apply the basic color that appears in the top row of the color palette, then simply set the percentage figure to 0 – like this:

```
{"solid": {"color": {"expr":{"ThemeDataColor":{"ColorId": 6, "Percent": 0}}}}}
```

Colors based on the color palette can be applied after several possible keywords:

- Color
- Background
- Title font color
- Background font color
- Value font color
- Indeed, anywhere a color is set in the JSON theme file

However, cascading color references *cannot* be applied to the simple color definitions such as text classes and elementary colors (*firstLevelElements*, etc.).

Interestingly, in the actual JSON, you can define *any percentage variant* of the initial theme color that you like and do not have to stick to the percentages used by the color palette. So you could, for instance, define a variant of a theme color like this:

```
{"solid": {"color": {"expr":{"ThemeDataColor":{"ColorId": 5, "Percent": -0.37}}}}}
```

This allows you a wide range of potential shades based on the original palette.

As an example of cascading colors, take a look at the following JSON code. This can be found in the file CascadingColors.json.

```
{
"name": "Cascading Colors",
"dataColors": ["#12239E", "#118DFF", "#00FF29", "#6B007B", "#FFB000", "#F5F4F0", "#FFFF00",
"#FF001E"],
"visualStyles":
  {"*":
    {"*":
      {
      "total":
        [{
        "totals": true,
        "label": "TOTAL",
        "fontColor": {"solid": {"color": {"expr":{"ThemeDataColor":{"ColorId": 9,
        "Percent": -0.5}}}}},
        "fontFamily": "Arial",
        "fontSize": 10
        }]
      }
    },
  "tableEx":
    {"*":
      {
      "total":
        [{
        "totals": true,
        "label": "TOTAL",
        "fontColor": {"solid": {"color": {"expr":{"ThemeDataColor":{"ColorId": 8,
        "Percent": 0.6}}}}},
        "backColor": {"solid": {"color": {"expr":{"ThemeDataColor":{"ColorId": 9,
        "Percent": 0.35}}}}},
        "outline": "Frame",
        "fontFamily": "Arial",
        "fontSize": 12
        }]
      }
    }
  }
}
```

In this JSON

- The font color for the generic total font color is set to 50% darker than the final (rightmost) column in the color palette. This corresponds to the bottom row in the palette.

- The font color for the table total font color is set to 60% lighter than the final (rightmost) column initial color in the color palette. This corresponds to the second row in the palette.

- The font color for the table total background color is set to 35% lighter than the penultimate column's initial color in the color palette. This is a custom shade that is not directly available in the palette.

Once you have set up colors in this way in a theme file, it becomes ridiculously easy to change the colors throughout the theme. All you have to do is to modify the colors in the initial definition of the eight palette colors – and all the colors in the theme are updated to reflect the new choice of core colors.

Conclusion

You have learned in this chapter how to accelerate dashboard formatting by defining cascading style elements and color application. You learned how to define visual formatting using a mixture of high-level, more generally applicable stylistic definitions and then complete (or even override) these with tailored definitions for specific visuals. You also saw how attribute definition can be applied using a hierarchy of styles.

Then you saw how to create cascading color definitions that enable you to refer your color choices back to the color palette. This enables you to avoid hard-coding color choices and bring a new flexibility to your theme files.

With this, you have finished your tour of Power BI themes. I hope that you have enjoyed this book and that it will help you save time when creating Power BI dashboards.

■ ■ ■

Sample Theme Files

Sample Theme Files

If you wish to follow the examples used in this book – and I hope you will – you may need sample theme files to work with. All the files referenced in this book are available for download and can easily be installed on your local PC. This appendix explains where to obtain the sample theme files and how to install them.

Downloading the Sample Data

The sample files used in this book are currently available on the Apress site. You can access them as follows:

1. In your web browser, navigate to the following URL: github.com/apress/pro-power-bi-creation-2e.

2. Scroll down the page and click the tab Source Code/Downloads.

3. Click the link Download Now, and choose a directory where you will save the file PowerBiThemes.zip.

You will then need to extract the files and directories from the zip file. How you do this will depend on which software you are using to handle zipped files. If you are not using any third-party software, then one way to do this is

1. Create a directory named C:\PowerBiThemes.

2. In the Windows Explorer navigation pane, click the file PowerBiThemes.zip.

3. Select all the files and folders that it contains.

4. Copy them to the folder that you created in step 1.

Index

Printed in the United States
by Baker & Taylor Publisher Services